Psychology
from Conception to -

ISBN: 978-1-4092-7218-2

Psychology from Conception to Senility

Published and printed by:

Andreas Sofroniou

33, SN3 1PH, U.K.

Copyright   © Andreas Sofroniou, 2011.

ISBN: 978-1-4092-7218-2

FOR MOTHER

THESE FEW WORDS FOR MY MUM

WHO IS WORTH MORE THAN A HUG.

ON EARTH, IN THE WORLD SHE IS THE BEST.

MY GUARDIAN ANGEL.

YOU ARE MY MUM,

FOR EVER MY PAL,

GIVING ME WARMTH AND LOVE.

ALWAYS THERE, TO SEE TO MY NEEDS.

FOR EVER BY ME, VERY NEAR,

WISPERING COMFORT IN MY EAR.

YOUR WORDS OF BEAUTY, MY IDEALS,

DREAMS TO BE FULFILLED, SO REAL.

END.

ANDREAS SOFRONIOU'S ANTHOLOGY, ISBN: 0 9527253 0 4.

<u>CONTENTS</u>

	<u>PAGE</u>
POEM: FOR MOTHER	6
POEM: THE UP-KEEPERS	7
PREFACE	8
POEM: THE MOUNTAIN	11
1. INTRODUCTION	13
POEM: THE TRIO	16
2. CHILD CULTURE	17
POEM: THEANTHROPOS	27
3. THE PHYSICAL ASPECT	29
POEM: LITTLE BIRD	34
4. PSYCHOLOGICAL FACTORS	36
POEM: FORBIDDEN LAND	46
5. THE PREGNANCY PERIOD	51
POEM: MODESTY	53
6. CHILDREN'S IDEALS	54
POEM: THE CYPRESS MAN	59
7. HEREDITY AND THE PATERNAL INFLUENCE	60
POEM: THE SNAKE, ON YOUR BIRTHDAY	67
8. THE DAY OF BIRTH	69
POEM: THE OLIVE TREE	79
9. CARE FOR NEW-BORN CHILD	81
POEM: DESTINATION	87
10. THE POWER OF WORDS	88
POEM: MY TREE	93
11. ANSWERING QUESTIONS	95
POEM: PARADISE	104
12. OVERCOMING DIFFICULTIES	106
POEM: EROS	110
13. THE HEALTH OF THE CHILD	111
POEM: UPBRINGING	114
14. HABIT FORMATION AND PUNISHMENT	115
POEM: DILEMMA	121
15. PERIOD OF TRANSITION	122
POEM: ROUTINE	128
16. RESPONSIBILITIES REVIEWED	129
POEM: THE BETTER HALF	132
AUTHOR'S PROFILE	133
SYNOPSIS	134

THE UP-KEEPERS

LOOK AT THE YOUTH AND SEE THE FUTURE,

THINK OF THE PRESENT AND SENSE THE TOMORROW.

PICTURE THE LAND, IMAGINE THE FRAMEWORK,

DIG DEEP THE FOUNDATION OF TODAY'S YOUNG ONES.

PREPARE FOR THE BUILDING OF THE LAND OF NOW,

KNOW THAT THE YOUNG OF TODAY ARE THE COUNTRY OF TOMORROW.

ASK THE TEENAGERS WHAT THEY HOLD FOR THE FUTURE,

SEE THAT THEY HAVE NOW, FOR THE GENEROSITY OF LATER.

FEED THE EMBRYO WITH MUSIC, LOVE AND CERTAINTY,

HOLD THE TODDLER TIGHT AND BE THERE IN TIMES OF NEED,

CONTRIBUTE WITH OPEN ARMS TO THE YOUNG AND UNBORN.

REMEMBER TODAY'S YOUTH ARE THE WORLD OF THE MORROWS.

END

ANDREAS SOFRONIOU'S ANTHOLOGY, ISBN: 0 9527253 0 4.

PREFACE

It is believed that the subject of offering advice, guidance and in general, counselling on being a parent in modern times has not really been addressed. With this in mind, this book is written in a simple language. Although in plain enough English, the content will still be of importance to the reader and without any compromising on the training of the parents. There is no doubt that this subject is deep and vast. Backed by recent social events and political debating, the Joyful Parenting can only be helpful in preparing for a child and in bring up a family.

The book concentrates on the upbringing of children and offers guidance in establishing the right relationship between the child and the parents. It deals with the pre-natal and post-natal influences and expands into the realms of continuous development of the human personality. Remembering that human personality with all its complex characteristics never stops developing; from the foetus stage, to birth, growing up and to dying in old age.

If this book is to accomplish its purpose and be of real service to the mothers, fathers and the society of the future generation, it must be applicable and adaptable to people in all walks of life, simple and direct in its contents. It would be a comparatively easy task to write a treatise on Child Psychology, even easier for the author to write his experiences based on psychotherapy and counselling covering almost two generations of practising.

It is very easy to write for men and women already well advanced in the study of this subject and who have abundant means to supply the needs of a growing family. But, there are others in less favourable circumstances, those who may need instruction badly and any attempt to write on the psychology of the child must be broad enough to reach all who might be helped by such a book.

Many will understand the explanations which are given carefully and grasp quickly the subject which is treated at considerable length. These details are set forth so that no one who undertakes the reading of this book may fail to comprehend the thought intended and be benefited thereby.

The book is offered to all parents, teachers, leaders and to those who care about the future, the generations that will follow the present legacy. It is hoped that people in all walks of life will find the contents simple enough; a guide towards the understanding of the inter-relationship with the child.

THE MOUNTAIN

A MOUNTAIN AS HIGH AS THE SKY

STOOD IN FRONT OF ME,

IN MY WAY STILL - QUIET

AND I TALKLESS, MOTIONLESS

AND IT, HARD AND ROCKY.

I TRIED TO MOVE,

I WISHED IT WOULD MOVE;

FORWARDS

TO IT I TRIED TO GO

BEGGED TO ME, COME.

I CRIED FOR THE MOON

THAT DIDN'T EXIST,

I WISHED FOR THE MOUNTAIN

TO MOVE THAT COULDN'T MOVE, I SHOUTED

AND THERE WAS NO VOICE IN ME.

I THOUGHT OF CLIMBING

AND THERE WAS NO STRENGTH IN ME.

I THOUGHT OF THE WIND,

IF IT BLEW..,

THERE WAS NO LIFE.

THE MOUNTAIN

STOPPED IT ON THE OTHER SIDE.

DRY, ROCKY AND STILL,

LIKE THE MOUNTAIN

THE VALLEY WAS.

EVERYTHING DEAD

ALL TO AN END

MOUNTAIN-LIKE

DEAD - STILL.

END

ANDREAS SOFRONIOU'S ANTHOLOGY, ISBN: 0 9527253 0 4.

1. INTRODUCTION

In raising a child, there is no reason why prospective parents should be reminded of the simple rule of love and care for their offspring. This being a basic need, it is taken for granted that everybody is well equipped for this responsibility. Loving a child is one of the first and foremost of human motives. In most of us, it comes natural. We wish for the conception, we want the best for the child and we crave to hold the little one in our arms. Many feelings go with love, affection, warmth, and positive thoughts. From the moment of conceiving, all the comforts of the pre-natal stage in the mother's womb and the early stages that follow the birth.

Often forgotten or neglected, is the father's positive contribution and feelings that accompany the mother's love for the unborn child. The child's experiences in the womb and the subsequent post-natal caring form the basis and the beginning of a great loving relationship between the child and the parents.

The pre-natal stage and the post-natal understanding of the child's needs, help enormously in the formation of the child's character. At this stage, the development of the child's personality, his/her education, the understanding of values, all this become predominant in the parental home environment. This requires the contribution and the effort of both parents and whenever, if possible, the attention of the extended family.

Such care for the unborn or newly-born child requires a little bit of planning and preparation for the big event of giving birth and of preparing the child to enter a chosen society. It cannot be disputed that the growth and the upbringing begins in the mother's womb. Even less arguable, is the fact that the parents need maturity, knowledge, understanding, training, and guidance in undertaking such an important responsibility. Social demands on the child and the parents are such that some reminding to rearing a child becomes a necessity.

Parents should not feel guilty whenever they seek help in understanding the child and his/her upbringing. Such an action should be considered wise. Guidance towards parenting should be readily available, in clinics, in surgeries, in counselling establishments, in book-shops, in academia, on book-shelves and on the coffee table.

There is no age limit in asking questions and in enquiring as to the best method for caring for your child. The reward for loving and caring for a human being is the most satisfying feeling ever described. To feel that you have done your best for that child is a personal sensation that nothing else can compete.

To perform well as a parent it may call for some form of guidance. If it means reading on how to give your love and offer your care to your child, so be it. There is no reason for holding back your love and care and certainly no room for quilt regarding your reading on such an important topic. Parents spend a great part of their lives working and preparing for a family. Any opportunity given in parental education ought to be pursued.

THE TRIO

THE SUN THAT LIGHTS,

THE FLOWERS THAT BLOOM,

ARE NATURE'S GREATEST BOOM.

NATURE NEEDING

THE SUN AND FLOWERS.

WHAT CAN THE FLOWERS DO

WITHOUT THE SUN'S BRIGHT LIGHT

AND WHAT WOULD THE SUN'S PLEASURE BE

WITH NO FLOWERS UPON TO BEAM?

BUT HOW CAN NATURE PREVAIL

WITH NO FLOWERS AND SUN BEAMING ON IT?

A LOVELY TRIO,

UNIVERSAL MIND'S CREATION.

STRONGLY AND EVERLASTING ARE BOUND

TO LIVE THE ROUND

OF INFINITY AND DESTINY, TOGETHER.

 END

ANDREAS SOFRONIOU'S ANTHOLOGY, ISBN: 0 9527253 0 4.

2. CHILD CULTURE

The human race, in its evolution, has made a very slow progress. All the way it has had to make experiments in many ways, sometimes succeeding and taking a step forward and on many occasions experiencing failure and re-trying. Where people have gained by their mistakes, failures have in a way resulted in progress, but where failure has not been overcome, the same mistakes have recurred.

Political and other leaders, mainly in education and religion, have been advocating for many years that the cause underlying most of the difficulties human beings experience as adults, is the wrong training received in early childhood. While the leaders' claims appear reasonable, little improvement in the things which make life most worthwhile is apparent as the result of all kinds of efforts.

Achievements in all forms must be judged by the benefits which they bring to the human race and if the methods of child upbringing are not producing a greater measure of freedom, prosperity and happiness throughout life, then something must be wrong with the system of educating and caring for the child.

Therefore, the system of child training must be reviewed and examined at every step, to find where the failure lies. As the popular education is looked into, there is much that is commendable and perhaps a few radical changes could be made without altering the entire method of educating society. It may become obvious that the early years of the child are the vital ones.

It may appear logical that any failures cannot be attributed to the child and that advising counselling and educating may not be for the child, but rather for the parent, the guardian, the teacher, the adult. For in their hands lies the future of the child, the future of the human race, the next generation.

Psychology, some say is the science of human behaviour, or the study of the mind and out of the psyche in its various sections, grow all the feelings and emotions. All actions and conduct of the individual depend on sentiments - feelings and emotions.

Whatever the pattern of behaviour, whether from a mental, or material point of view, the attitude of the individual is determined by the action of the conscious mind and its reaction upon the subconscious.

Astonishingly, although it is well known that the mind is behind all actions, little attention is given to the study of the mind and its relation to life in general and the environment.

If a person wants to learn about agriculture, he/she is expected to know about the soil and the best method to yield the maximum possible. A car mechanic must know everything about the car and he/she is expected to undergo the appropriate apprenticeship and training and subsequently gain a lot of experience in the field. But, people in general give little, if any attention to the functions of the psyche, which they must use during the whole of the existence on earth. The skill in the use of the mind determines the success or failure in life.

Whether a farmer, a computer scientist, a teacher, or a labourer, the mind is behind any actions and in performing one's own duties. Hence, the importance of the study of behaviour and human sciences must be remembered. Just as important, being the relationship of Homo sapiens (being the major intelligent manipulators and participators) to the whole of the environment. Not forgetting the effect of the human actions to the surroundings and the influence on what the next generation will inherit.

In helping a child to grow, one must realise that whatever it is or does, desirable or not, is the result of the mental action. Since the action of the mind of the child is influenced by others, it must be realised how important the behaviour of those close is. Predominant the influences can be from the mother, father, and siblings. Extending to teachers and everybody else coming into contact with the child.

Each generation has two distinct tasks to perform. One is the responsibility to advance the present civilisation to the highest possible point. The second task, a simultaneous one, is to prepare the generation that follows; equipping the young ones with the necessary skills, respect, knowledge and the understanding of life at large. The outcome of this is the continuation of civilisation onto a higher plane.

Engrossing one's interest on of the two responsibilities is not enough. There must be equilibrium between the two tasks. History presents many examples of civilisations that became so obsessed with arts, sciences, conquests... to the point of neglecting the training and care of the children. When children, thus neglected, get their chance they make a shipwreck of the fine achievements for which the previous generation gave everything.

Many neglected children have a tragedy to compose. The father may be a successful businessman and the mother outstandingly successful in politics, but the time comes when the children cannot inherit the full values of life. Inheriting the material successes of the parents is not enough.

Many times, these children enter an era of dissipation, squandering not only the accumulations of their parents, but much of what other generations had acquired. Such parents defeat their own purposes in rearing children who lack the ability to continue the work and ideals of the family. Parents may glory in their success and point with pride to their achievements, but as long as the children are not taught the sanctity of life and the importance of individual responsibility, little progress can be made.

The subject of child education and development is often mentioned by leaders and the mass media, but does not appear to be of sufficient importance by the law-making bodies. Certain authorities believe that it is very difficult to secure enough funds for schools and education. Teachers are earnest and sincere in their efforts for the children at school, but are handicapped by crowded conditions, lack of equipment and in a few cases undernourished condition of the children.

The cutting of school funds and teachers' salaries is something much more serious than its immediate effect. It means that the educational system and the human resources are morally low and de-motivated and as such the children, the coming generation, will not be a worthy successor of the national inheritance. It is well known that parents would rather suffer hunger than have the children do without. But legislatures approach the all important subject of child education without any realisation of the momentous consequences which they control.

The ancient Athenians, with all their wonderful ideas of beauty and equally outstanding skill in objectifying, they failed to impart these powers to their successors, the younger and next generation. With such certainty, one can say that this was the major contributor the end of the Hellenic glory. Every generation is jeopardised by the danger of entering so enthusiastically into its achievements as to neglect its childhood. A neglected childhood is always ready to make junk of the achievements of its earlier generations, the arts, sciences, technology, inventions, even of their dreams.

The child is a part of the human race and is attached to the parents. To attempt to understand the child as a unit by himself/herself, without taking into consideration all the influences brought to bear upon her/his life by parents, teachers, religious leaders and others, would be as great a mistake as to try to comprehend the value of a car by looking at the shape alone. The support of the child would not be complete if attention is only given to the appearance of the child. If the child is to have the finest of development, in its holistic sense, those who bring influence must be considered, especially the mother.

There is a very strong bond between the mother and her child. Her influence upon the child before birth is great and reaches back, even more, into the life and character of the mother. This primary influence of the mother, together with many other factors having a part in the early life of the child, must all be taken into consideration in bringing up a child.

Carefully regulated habits and a high standard of conduct for the child are not sufficient to prepare him/her for life. These are of value, but they are only the outer shell which contains the major characteristics of the child.

The different periods of development must (each stage of development) be given the consideration it deserves. Psychologists divide the development of human beings into parts. Simply speaking, the influence of the mother and others on the individual expression of the child prior to birth and until emotionally grown. This is known as the pre-natal period of growth. The second part of development is the post-natal stage up to toddler-hood. During this period, there is a tremendous influence of the mother upon the child, particularly if the child is breast-fed. These first years are of inexpressible importance in the development and formalisation of the character of the child.

The third part of the child's development extends into adolescence. The child's main functions are based on earlier experiences and the influences of the present surroundings. This is the time for the parents and those near who are interested in the welfare of the child to sow seed which they wish to see come fruitage later, for whatever is planted in the soil of the child's mind in these early years is sure to bear fruit.

Later in this period the child passes through a time of great importance in which physically and mentally he/she is being prepared for a change which becomes manifest in the adolescent period. If understanding and tact are used in the direction of the child through these years, the child will be well-prepared for the next stage of his/her journey toward maturity. Many questions are in the mind of the child as he passes from early childhood into a larger world where there are many things difficult to understand. Much wisdom, great love and patience are required for the safe passage of the child through these years.

During adolescence, care must be exercised to see that no harm comes to the child while she/he is bringing the gap from childhood to manhood or womanhood. These are precious years and to fulfil their purpose will require consecrated effort and interest on the part of the parents and teachers. It is sometimes called the awkward age and the child is laughed at because he/she cannot gracefully accomplish all the re-adjustments that life seems to demand at this time. During this period children are particularly sensitive and their pride is easily wounded. They may become sensitive and uncommunicative when in the presence of grown-ups. Empathy and understanding, based on one's ability to remember her/his own youth, will form a bond between adult and child at this time which the years cannot break.

The next phase of development is the ushering of the youth into maturity. As the child grows older, the parent, teachers and others responsible to a degree for her/him, must learn the ratio of their diminishing responsibility, which ends completely when the child arrives at absolute accountability for reaching his/her own conclusions.

THEANTHROPOS

I CAME OUT OF THE WOODS TO HUNT IN THE FREEZING WIND,

I SAW THE FURRY ANIMALS; I KILLED ONE TO EAT,

USED ITS HIDE TO KEEP WARM, CONTENTLY SAT UNDER THE
TREE

IN TROPICAL RAIN, NOTICED THE SNAIL PROTECTED IN ITS
SHELL.

IN ENVY I CUT THE TREES AND BUILT A HUT TO PROTECT ME.

SO CONFIDENT I BECAME, CLOSELY I IMITATED NATURE -

I TRAVEL IN LOCOMOTIVES, AUTOMOTIVES, AND THE LIKE.

I LAYED DOWN IN THE BLISS OF SECURITY LOOKING HIGH
ABOVE

AND OBSERVED THE BIRDS GLIDING AND FLYING AWAY TO
PLACES.

ERECTING MYSELF I STARTED MIMICKING THE WINGS -

SLOWLY BUT STEADILY BUILT THE PLANE AND FLEW BY THE
SEA.

IN DEEP WATERS, BY TRYING TO CATCH THE FISH, I GOT
SCARED

I COULD NOT SWIM FOR LONG, SO I CONSTRUCTED A SUBMARINE.

IN EMERGING, RESTLESSLY I FELT I WANTED TO REACH THE STARS.

I CONSTRUCTED A ROCKET AND WALKED ON THE MOON.

IN BED, FEEBLE AND MOTIONLESS, I LOOK AT THE CEILING

AND WONDER ABOUT MY INSIDES, I MODEL MY CELLS,

SPIRAL MY DNA AND ENGINEER MY GENES.

I PROGRESS IN TECHNOLOGY THAT HELPS ME TO RECOVER.

I GET UP AND IN VITRO I FERTILISE BABIES AND ORGANS.

NOW FERVENTLY, MY TECHNOLOGICAL MARVELS, I DECLARE -

I WANT TO CREATE LIFE, NO MORE A MERE MAN,

BUT A LITTLE GOD AND THEN PERHAPS, GOD!

END

ANDREAS SOFRONIOU'S ANTHOLOGY, ISBN: 0 9527253 0 4.

3. The Physical Aspect

It is claimed by some religious leaders that if the child is given into their care exclusively until the age of five years, they can absolutely determine the convictions of that child and, to a large extend, his character for the rest of his life. Where this claim may bear some truth, there is more vital a period, even before the first five years.

The most vital period in the life of the child covers the nine glorious months in which the mother carries the precious life within her womb. This is the mother's reign, a time when she rules absolutely in her kingdom. It is true she has but one subject, hers to make or mar. Mothers may not realise the power that is theirs during pregnancy, the veneration and awe, pregnancy being the greatest experience of life.

On the other hand, pregnancy is often accepted as an unavoidable evil, the child undesired and the mother looks upon this period as one of great sacrifice and hardship. No plans are made for the welfare of the little stranger, no high ideals held for the unborn child and no deep yearnings or mighty aspirations stir the mother's heart in reverent meditation. Nurtured in loving thought, rejoiced over in song, the child cannot but be great.

The hand that rocks the cradle rules the world, but most of the work has been done before the child arrives to occupy the cradle. The vital period is before birth. The mothers must realise their power and exercise it for the good of the child.

Pre-natal influence is now an accepted fact. In the past, hideous sights witnessed by the mother during pregnancy, or tragic experiences of the mother at that time, were blamed for birth marks and other abnormal conditions appearing in the child at birth. This is seeing only the negative side of the picture. It is true that there have been many individual cases where the mothers understood something of positive thinking and exercised it for the good of her unborn child with marvellous results.

The question may be asked, why this tremendous influence of the mother upon the unborn child? It is generally accepted that what is known as the subconscious mind carries on all the functions of the body. Therefore, the growth and development of the child within the uterus is under the direction of the subconscious mind of the expectant mother. This must be true, for the child has physical connection with the mother and this connection continues until severed after birth.

When the child is born we have a continuation of life which was. Hence, the great importance of a proper direction of the feelings, hopes, ambitions, inspirations and ideals of the mother. To a large extent, in the beginning of the child's life, its subconscious actions will be but a reflection of the subconscious actions and reactions of the mother.

One of the great assets of life is good health, a fact which does not need to be explained or enlarged upon. Every mother desires for a child a strong, healthy body, free from defects and with strong power of resistance.

The expectant mother may not herself possess vigorous health. She may be subject to various forms of ailments and lack strength and vitality. When a woman of this type realises that she is pregnant, she should at once begin to picture in her mind the kind of child she wants. She should see in her mind a strong, healthy child, romping and playing, never ill, always happy and full of life. This mother-to-be should minimise her own aches and pains and magnify and enlarge upon any short or long periods of physical well-being which she may enjoy. She should not listen to tales of sickness or accidents and, so far as is possible, keep her mind in a happy, peaceful state.

If possible, she should associate with strong and vigorous people whose tone is uplifting, because its key-note is health and laughter. She should not be afraid of fun and jollity, for laughter is a splendid physical tonic. The pregnant woman cannot afford to jeopardise the well-being of her child by giving attention to anything but the very highest and best within her reach.

She should look at pictures of handsome men and beautiful women whose faces express strong character, for she wants her child to be endowed with good looks as well as strength and health. Beauty of face and body in an individual represents a higher expression of self from the physical standpoint. So, in her dreams of the child-to-be, let the mother build beauty, strength and health.

This reading will not be complete without a few suggestions on diet and personal hygiene. The mother will see that her menu contains plenty of fruits and vegetables, for they are the foods which contain the valuable mineral elements which are so important in building a strong, symmetrical body. She should drink plenty of water.

Exercising, such as walking, or any other light exercise should be taken daily. Housework, if it is not too heavy and exhausting, is good, but the expectant mother should spend some time outdoors each day. She should take a bath frequently, everyday if possible and take a good rub after leaving the bath. The teeth should be brushed three times a day, or after every meal. Instinctive cleanliness is a thing that is passed on to the child. Man is a water animal; the child in the womb is formed in fluid and lives in fluid for nine months. It is born in water and if the mother likes her bath and relaxes and rests in the water, the child will be much easier to train in taking its bath. Some babies tease for their baths, even before they can sit up, when they see the water being prepared; other babies not only seem to dislike water in general, but kick and fight and almost go into convulsions at bathing time.

The pregnant mother should have plenty of rest and sleep, sleeping at least eight hours out of the twenty-four. She should guard against getting over-tired and exhausted. If she is engaged in housework or about other matters, she should rest several short periods a day. Have a siesta.

LITTLE BIRD

Tiny little bird with the weak chirp

And the boastful colourful chest

You are growing by the day with grace

With huge effort learning to flap the wings.

Hopping instead from branch to branch

Falling on twigs and with fear holding on

Feeling the pain on your small claws

With your pride hurt, straightening up.

The fleeting mosquito you try to catch

Your beak opening wider than needed

The insect escaping, still hungry

In desperation, you make a sudden move.

Falling down is another lesson, scared

You are on the ground, unfamiliar growth

The crawling worm frightens you more

You hop and fall again, famished.

Your last hope is your thought, the familiar

The mother with the nourishing mouthfuls,

With loud chirps she calls you, feeding time

Makes you run and flap to her, safety at last.

END.

ANDREAS SOFRONIOU'S ANTHOLOGY, ISBN: 0 9527253 0 4.

4. The Psychological Factors

It is a difficult matter, to discuss the physical body and the mental life separately, for man is so constituted, that he is a unit of body and mind. Every phase of his being overlaps and inter-wines the other phases. It can be said that the mind influences and controls the body and the body in return reacts upon the mind favourably or unfavourably as the case may be and may determine the next mental action. Hence, many of the statements regarding the physical well-being will apply to mental culture too.

It is important that every precaution be taken, so that the child may come into the world with a perfect physical body. If we were to stop preparing for the physical well-being of the child, we would fail far short of accomplishing our task. The mind of the unborn child is of equal, if not of greater importance than its physical body.

Since the function of the mind is to think, we must start with the mother's thought and endeavour to help her to keep her thinking running in the right channels. It is not the idea that the mother shall become a logician, delving into deep philosophical or scientific questions, but she must learn to control her thoughts and train herself to think the kind of thoughts which will produce desirable results in her child. For thoughts are creative and produce after their kind.

The expectant mother must free her mind from all thoughts of so-called evil, such as criticism, jealousy, envy, unkindness, cruelty, gossip, fear, anger, greed, misery and selfishness. It cannot be done by simply trying not to thing of theses things; it can only be done by substituting in their place the best thoughts of which she is capable - thoughts of love, kindness, peace, courage, strength, hospitality, forgiveness beauty and goodness. The list is endless, but this is not easy. For when one is stirred by incidents which ordinarily produce irritation and anger, it is difficult to hold the mind on thoughts which represent the opposite of the way one is feeling at the time. However, it can be done and must be done if the mother is to become an instrument for good to her unborn child.

One means of accomplishing this, is for the mother to memorise some lines of poetry or prose which express beautiful thoughts, poetry and ideals, or which are simply statements of truth. These may be just one or two lines, but if they are memorised they can be used at any time.

Sometimes, feelings seem to be stronger than at other times and require all one's energy and strategy to eliminate them. If she has some beautiful thoughts memorised, she can repeat them over and over, compelling her thought to dwell upon them. The repetition of the lines can continue until finally the feeling of irritation will be forgotten for the time being and the mind will become quiet and peaceful.

It is true that the discordant feeling may return and then the work must be done again, until finally the enemy is vanquished and harmony is restored. For discordant thoughts are the worst enemy one may have and for the expectant mother to entertain such thoughts, means the destruction of any high ideals she may hold for her unborn child.

The expectant mother should read books on biography and travel and try to remember the happiest, finest, noblest incidents. When some ugly thoughts flash into her mind, as they may have a habit of doing, at once she should turn her mind on some of the pleasing pictures which she has been storing up in her mind for just such an emergency.

The power to think is so tremendous that it may well be given earnest and serious attention by any man or woman who expects to build a character worthwhile. But in its importance to the expectant mother, it outweighs every other factor in her preparation for the coming child. It is not easy to control one's thinking and produce thoughts that are of a constructive nature. It is a task worthy of the best mettle, but its rewards are commensurate with its difficulty of attainment. While the mother may not be assured that she will succeed perfectly in weeding out all undesirable thoughts and substituting only the good, nevertheless, she owes it to her child to do her best.

She must deliberately decide what she will think about. If she permits herself to think thoughts that are not constructive, she is not protecting her child, but is blasting its future with the results which her thoughts are sure to produce. Many-a-mother would willingly lay down her life to save her child from the results of its folly, when in its youth or early maturity it has become involved in serious difficulties.. These difficulties might have been avoided had the mother understood the law of pre-natal culture and practised it in her thought life, during the time she was carrying her child.

A woman's thought during pregnancy will naturally turn often upon herself, her present condition and the time when the child will be born. During this entire period she is hyper-sensitive, peculiarly susceptible to her own suggestions and the suggestions of others. So it is comparatively easy for her to impress upon herself the ideals which she covets for her child and along this line her thoughts should be directed.

If there is any fear or anxiety for her safety at the time of child birth, she should get rid of it at once. Pregnancy is natural, there is nothing to be feared, for women have been giving birth to children through the ages. Woman has been prepared for this service and her body formed as it is for this particular purpose. At this time, if she will suggest to herself strength, courage and high purpose, it will have a splendid effect upon herself and also the child. The percentage of cases where there is serious trouble at child birth is much lower than we have been let to believe and in many such cases, this trouble could have been lessened greatly if not avoided altogether, by the practice on the part of the mother of the principles set forth in this text.

The mother should be careful to entertain no thoughts or feelings of self-consciousness or timidity. There may be a tendency toward embarrassment because of her physical appearance, particularly through the last few months of pregnancy, but there should be no sense of shame or embarrassments, no more than there should be in the fruit tree because of the enlargement of the bud for the coming forth of the flower and fruit. If there is any such tendency in her feelings, she should overcome it and force herself to socialise, meet and talk with people.

She should not permit her appearance to keep her within doors and away from people. She should go out and mix with others. But, be sure that her associates are persons who have high ideals and the right attitude toward life.

The mother-to-be should attend lectures, concerts, social events, church services and any entertainment of high order. If a woman remains in solitude during pregnancy, her child is likely to be timid, fearful, backward, and retiring. Therefore, for her child's sake, the expectant mother must continue her social life and keep up her interest in the outside world.

No great attempt should be made to wear clothing that will conceal the fact of pregnancy, but rather the mother should be proud of the fact that she appears to be in preparation for the accomplishment of one of the highest and noblest expressions of nature.

She should not allow anyone to express sympathy for her because she is pregnant, or feel any regrets in the matter herself. Rather she should glory in this wonderful experience and make it count to the very utmost in her own life and in the life of coming child.

This is not the time for the expectant mother to become mentally lazy and inactive. Additional to her physical exercises, the mother should exercise the mind, as well. She should keep her mind busy along constructiveness of thinking. She should read good books and think seriously for at least a short period each day on some statements. Poems, or ideas expressed in her reading. If she keeps her mind occupied with the highest thoughts, trying to understand things and philosophical aspects, her power to think will increase and she will advance step by step in her mental development. This will increase the mental capacity of the coming child.

Regardless of appearances, there is nothing to worry about. She should think the brightest, happiest, most kindly loving thoughts she can muster into her mental realm.

By now it has been shown that while the mother's thoughts have a direct bearing on the child's life, it also has a very important influence upon her own life, both present and future.

The usual feeling of the average parents toward the coming child is one of pleasurable anticipation of having a child in the home to love and care for. They may have high hopes and ambitions for that child, but rarely do parents expect their child to go far beyond their own attainments. By their own thinking regarding the child and their low ideals for him/her, before birth, an influence which is likely to limit and narrow the child's accomplishments.

Sometimes, a woman becomes pregnant against her wishes and there may be a tendency to rebel and there are cases where a mother entertains a feeling of resentment. For her own sake, as well as that of the child, a woman in such a situation will do well to consider carefully what she is doing and bring her thoughts into line. It must be realised that the child is not responsible for his coming, that he/she is an innocent participant. The mother, by her thoughts can make this child a blessing to herself. On the other hand, by the terrific influence which she is able to bring on the growing embryo by her strong feelings of hatred, she may brand the child and doom it to a life of crime.

So, if conception is not desired, every feeling of displeasure or regret should be put aside and feelings of the opposite nature should start. It is the woman's task to find these happier pictures and build into them a mother's love and joyous expectancy and welcome.

The expectant mother should see her child perfect, happy and strong. What she expects comes to pass, through positive expectation and thought. This attitude of right thinking is all important in giving the mind of the child a good and happy impression. Avoid as far as possible, arguments, emotional scenes and anything that will cause her distress.

The expectant mother should avoid and reject evil thoughts as she would poison. She must be careful of her feelings which are the result of her predominant state of mind. As she cultivates thoughts of beauty, love and peace, these attributes will gradually become her predominant feelings and will produce, in her life and the life of her child, pleasing and happy results.

A happy relationship should be established between mother and child during the nine months of great intimacy. In future years, the mother will appreciate the companionship and confidence of her offspring.

The time to establish this close relationship, which will last throughout life, is while the mother is carrying her child. She should in mind talk to her child of her own highest ideals and any secret ambitions she may have had which were not fulfilled. She must not try to impress these ambitions upon her child for their fulfilment, but talk it over with the child as she would with a most loving and understanding friend. Then she should talk to the child of the great possibilities before it of the wondrous beauty and grandeur of life.

The mother should see her child coming to her with all of its little trials and difficulties, knowing that the mother will always be ready with her sympathy and help. In this manner a mother may prepare the way for a very close and happy companionship which will be a lifelong asset to both.

In thinking of her child in such a manner, the mother will see it playing with other children and associating with those of its own age as it grow in years. The mother's place cannot be filled by another, neither can she try to fill the whole of the child's life. She is only one of the many friends her child will require for a well-rounded and complete life. Above all, she should not bind the child to herself to the exclusion of others.

The mother should not be jealous of the love of her child. However, much the child may come to love others, if she is of the right mental attitude, there can be no question of the child's love for her.

The mother should suggest to her unborn child, an eagerness for knowledge. Should see it excel in education and with many interest and talents in life and with many accomplishments.

THE FORBIDDEN LAND

HIGH ABOVE THE TREES

TWO HAWKS ARE GLIDING.

BIRDS OF PREY, SPYING

THEIR EYES, FIXED TO THE GROUND.

THE HAWKS' SCREECHING NOISES,

EAR PIERCING TO THE LITTLE ANIMALS,

BRING FREEZING FEAR AND WITH COLD SWEAT

BIG AND SMALL CREATURES RUN TO HIDE.

WITH THEIR TAILS IN THEIR HIDES

AS FAST AS ONE CAN GO, THEY CRAWL

HASTILY INTO THEIR HOLES, UNDERGROUND,

CLOSE TO EACH OTHER, THEY FEEL SAFE.

HIGH ABOVE IN THE SKY, NEAR THE CLOUDS,

THE BIG HAWK WITH AUTHORITY, TALKS

AND THE SMALLER HAWK RESPECTFULLY LISTENS

TO THE MYTHS OF TIMES PASSED.

WITH BRIGHTER DAYS YET TO COME,

THE TWO HAWKS CARRY ON FLYING

OVER NEW HILLS AND THE GREEN VALLEYS,

EXPLORING, NEW, AND FAMILIAR HUNTING GROUNDS.

THE OLDER HAWK SLOWS DOWN, HOVERING,

THE LITTLE BIRD COPYING, THE FATHERING HAWK

STRONGLY AND THUNDER LIKE EXPRESSES HIS FURY:

'AHEAD OF YOU IS THE FORBIDDEN LAND!'

THE FORBIDDEN LAND, THE HUMANS CALL

'HI-VAY AND MOTO-VAY, THE GIANT SNAKE

THE EVER-GROWING SERPENT, WITH COILS

THAT GLOW BRIGHTER IN THE DARK, CALLED CARS.

DAY BY DAY THE 'MOTO-VAY' GROWS LONGER,

THE COILS BECOME THICKER, OVER THE HILLS

THROUGH THE WOODS, THE SNAKE CRAWLS

WITH FUMES DARKER AND SMELLIER.

THE YOUNGER HAWK NOTICES

THE OLDER HAWK'S EYES, SADDENED

FOR THE FIRST TIME, WITH FEAR

INTO THE HIGH SKY, ESCAPING.

As the little one follows, into the clouds,

The father quietly utters 'that is the one, the snake

That took your mother away, one night'. A mother

The little hawk still in the nest, could not

remember.

The mother dived food to fetch, to feed

The newly hatched vulture, still in its warm nest,

Hungry, up high on the crag of the mountain, waiting

For his mother some food to scavenge.

The snake's moving coils, the humans call cars

On the back of the huge serpent, humans call moto-vay,

With the bright lights emitting, blinded the hawk

Broken wings, the female hawk flattened to the ground.

THE LITTLE MALE HAWK REMEMBERED, WITH SORROW,

OTHER SMALL VULTURES HAD MOTHERS, FEEDING THEM

AND HE HAD A FATHER, TEACHING HIM THE MYTH

ABOUT THE SERPENT, THE HAWK AND THE FORBIDDEN LAND.

END

ANDREAS SOFRONIOU'S ANTHOLOGY, ISBN: 0 9527253 0 4.

5. THE PREGNANCY PERIOD

If there is a picture in the home that is inspiring, the mother should spent time before it each day and visualise the ideals of which she is capable. Copies of the great masterpieces cannot be expensive and will serve this purpose admirably. Visit the art galleries and let the eyes and the soul feed on the marvel of art.

If there are flowers in the garden or home, she should spend as much time as she can with them, for they are beautiful expressions of nature. Also, frequent visits to the parks, town gardens, to the country and anywhere where nature speaks of peace and harmony.

The home environment should become as attractive as possible. Even if the dwelling is humble, it can be made neat and attractive, comfortable to live in, indeed a home.

The mother's personal appearance should be immaculate, wearing pretty clothes with bright, harmonising colours. This is no time for her to neglect herself. Now is the time for her to establish habits of which she will approve in the future.

In her visits to public places, the mother should listen to lectures on subjects of interest. She should go to theatres and cinemas. Many subjects are elevating in thoughts. Try and read some good literature, listen to soothing music and watch some good television programmes.

A prospective mother must make her words count for good. She cannot afford to become irritated and speak angrily to anyone. Kind, loving words should be used and should acknowledge any service rendered in appreciative words. In her conversation, she should talk of things which are of value.

MODESTY

The inanimate paper,

Which talk-less awaits

For black ink

And hand to be ensouled,

Made my happiness so great

When it told me

That it never expected me

With my endless verses

To write the sentiments of soul

And through the power of thinking

Such emotions to express.

But my modesty jumped up

And protestingly answered

That my narrow mind is not enough

To give it such a spirit,

Nor can I write just like that on it,

The already written paper.

END

ANDREAS SOFRONIOU'S ANTHOLOGY, ISBN: 0 9527253 0 4.

6. CHILDREN'S IDEALS

By now, so much has been explained about the mother's unlimited influence on the unborn child and the tremendous power which she wields during this period. A prospective mother can make of her child almost anything she desires. The mother may begin to plan the future of her child in detail, his profession, social standing, education, ideals, perhaps including some of her own ambitions, or some ideals which she may long have cherished.

Having set forth the mother's power along this line, it must be stressed that she does not own the child. The child is an individual in her/his own right and therefore not of possession or ownership to do with ass she pleases. It is hers to bring into the world under the best possible influences, hers to love, nurture and direct through childhood and youth, but here her rights in this matter end.

The child is an individual and no one has the right to choose for him/her the path which he/she will follow through life. That must be left for the child to do for himself. The mother has no right to impose her feelings and desires on the child. To do so, would probably make a misfit of the child and a failure of his/her life. Whereas, if she uses her influence to guide and encourage the child along the line of his own inclinations, he will most likely attain the goal which will represent his own wishes.

It must be remembered that the child being an individual, comes into the world with certain tendencies which will enable him to be more successful in some lines of endeavour than in others. Therefore, the education and training which the child receives, both before and after birth should be of the purpose of encouraging and developing these natural tendencies, for it will be along these lines that his greatest success and happiness will lie.

Hence, it is not wise, nor right for any mother to be specific as to the calling or profession or life work of the child, but the right mental attitude of the mother is to know that the child is good and successful in whatever he selects to do.

A mother with the best intentions may ruin the life of her child by choosing the vocation she/he is to follow. There are second rate doctors, lawyers and priests, who would have achieved marked success in other fields. The mother's ideals for the child should be the very highest, without being specific regarding the individualised expression of his life. She should want her child to have a fine physical body, beautiful in appearance, splendid health and individual ideals.

She should want him to be mentally alert, bright and eager to learn. But in her ideals for his education she will leave the details to be filled in by the child later on. In the matter of education, the world advances, technology becomes more complicated and educational systems change, so she should visualise the child equipped for life with the best education possible.

In the matter of the child's personality, she will want him/her to be with high ideals, honest and true in every sense of the word. This will necessitate high ideals for herself and integrity. It is important that she embody in her own life during pregnancy the virtues which she wishes to see depicted in the life of her child.

Spiritual values cannot be ignored in the life of an individual, in favour of wealth, power, and material achievement. In her thought of him, a mother will want him to be of great service to his fellowman, whatever line of work he may choose to follow. Success is to be desired, but it means more than the accumulation of money. A mother must include these in her daily periods of relaxation.

Practically, all study of Child Behaviour takes into consideration only the conditioning and the physical body of the human being, to the almost complete neglect of the spiritual world of the child, which is as vital and all important part of the human being for it determines what his is in life, his health, success and happiness.

A discussion of religion may be out of place in a book of this kind, but it is impossible to present any method or system of training for the child without an adequate understanding of this important truth, regarding his spiritual nature. It is not the intention of the writer to go into the matter of religion as such, for the mother may or may not be religious according to the usual conception of that term.

Nevertheless, everyone is more or less religiously inclined and the prospective mother will do well to develop her religious or spiritual nature during this period, even if it is for the child only. This does not mean that she will think of herself as a 'sinner' and count her numerous wrong doing. Avoid quilt or any fear for the future; emphatically no quilt feelings. Whatever the experience, is in the past and nothing can be done to change it. Make good use of the present, now.

In the quiet time of her day, she should think of some of the wonderful achievements of this age in science, invention, progress in technology and computing. Then try to realise that progress will continue, into the future. If the mother wishes her child to be the channel for high ideas to be used for the advancement and improvement of mankind, she will keep her thoughts as much as possible on progress.

It is not intended that the mother should be up in the clouds all the time, but if she will endeavour to see the good in everything, the everyday facts of life will assume a brighter aspect and she will be able to move through a day of distracting duties, calm and peaceful and with confidence in the future of her child and all things in his surroundings.

Again, it is urged that she avoids anxiety, worry and all such destructive thoughts. If she persists in positive thinking, she will be able to get rid of the undesirable thoughts, simply because they are receiving no attention and they finally disappear. As the mother relaxes and keeps her thoughts on love and other positive things, she will have fewer thoughts of fear or worry and will gain confidence, strength, and courage for the overcoming of all the difficulties in her life.

THE CYPRESS MAN

HE STOOD VERY TALL,

THROUGH THE DOORS HE HAD TO BEND.

ALWAYS LOOKED IN CONTROL,

STOOD STRAIGHT WITH PRIDE AND A BIG SMILE.

BROAD SHOULDERS,

BULGING MUSCLES

ON THICK BONES.

WITH AN UPTURNING MOUSTACHE,

HE TALKED,

THEY LISTENED,

OFFERED OPINION,

THANKFULLY RECEIVED.

WHEN HE LOOKED,

YOU FELT THE SHIVERS.

MOTHER CALLED HIM GRAND-DAD

AND I RECALL HIM AS BEING A GRAND MAN.

END.

ANDREAS SOFRONIOU'S ANTHOLOGY, ISBN: 0 9527253 0 4.

7. HEREDITY AND THE PATERNAL INFLUENCE

So often, there are undesirable traits of character in the near relatives of the parents, or there may be diseases of one kind or another back in the families of either or both the parents. The mother may have some fear that her child may inherit tendencies that will handicap the child in life. But she should have no such fear. She must remember that at all times she stands between her child and inherited influences, that if her mind is pure, healthy and her thoughts directed to the right channels, that no undesirable inherited influences can have any effect upon her child.

Heredity is looked upon as a most powerful influence in life. Parents cannot and must not accept the inevitable, giving up without a struggle, consigning their child to failure and disease. Medicine, with the help of modern technology has progressed enormously into the realms of genetic engineering. Consultants of all related professions are as eager to help, that parents must keep confident in whatever the modern scientist may suggest. Their thoughts are always optimistic.

In other words, with the help of the modern specialist and the mother's continuous positive thinking, heredity may have whatever influence the mother and the father permit it to have.

If you have any reason for worrying on hereditary influences on the expected child, seek advice from the medical specialists, as early as possible during the pregnancy and adopt all the psychological factors explained in this book. The mother's mind with the help of the prospective father can move mountains.

In the matter of pre-natal influence, the mother will use her mind and the laws of nature to produce in her child something more desirable than might be produced if no thought is given and no knowledge brought to bear upon the case.

So far it may appear that the mother only is responsible for the child. This is not so. The father as well as the mother, plays a tremendously important role in pre-natal influences. For whether or not the mother is aware of it, the father has had much to do in the creative work of bringing this child into the world and the mother at least unconsciously realises that the father sustains an important relationship to the results of parenthood.

Perhaps no one will have as much influence on mother's feelings, inspiration, and mental state during pregnancy as the father. His position then comes to be a very important one. What then, should be his attitude toward the mother of his child?

In the first place, his attitude must be one of the greatest possible considerations for the mother. It will be his duty to be very patient with any whim or undesirable state of mind which the mother may undergo. For during this period, her whole physical being is at high tension. She will naturally be subject to expression of the best as well as the worst in her nature.

The father may take as a matter of course the expression of the best, but be alarmed and resentful at the expression of less desirable traits, such as being easily offended and disposed to find fault. Unconsciously, there may be a little feeling of sacrifice in the mind of the mother, because of bringing this child into the world for her husband. In the deep privacy of their life, she may be disposed to rebuke him for his willingness to satisfy his own nature at the sacrifice of her peace of mind and anticipated suffering.

The father will be wise to overlook such disposition and on the contrary enlarge on the blessings of the opportunity to bring into the world a life and personality that may make its influence felt for good and thus bring honour, joy and satisfaction to the parents.

The father will make opportunities for the mother to see the most beautiful and desirable things. He will take her out for walks and rides and feel himself duty bound to be in the finest possible frame of mind, seeing and pointing out the desirable and beautiful in everything.

During this period he will be the perfect lover and sweetheart of his wife; never in the days of courtship did he give as much attention as during this period. He should be thoughtful enough to send her flowers and shower her with attention.

In being unusually attentive to his wife during this period, he must be very careful not to show anxiety, or allow her to feel that it is just as it should be. If the thoughts of the father and the mother are what they should be, there can but come into the world, a life prepared for great achievement.

Thoughts are real, whether we give any expression to them by word, look, or action. Our thoughts have an influence upon those with whom we are associated. Particularly, this is true in the relationship of husband and wife. Therefore, it is quite important that the father allow no thoughts, or feelings of anxiety to occupy his mind for if he does, the mother will register these thoughts of which she may not be and perhaps is not aware , but harm is done.

Hence, the father must think as kindly and encouragingly of his wife in her absence as when she is present. He must not discuss with his most intimate friend any undesirable feature of the matter. Because the father's character, life style and consciousness, registers upon the subconscious mind of the mother and therefore on the mind of the unborn child.

The father should guard carefully the physical well-being of the mother, seeing that she does not overwork, that she has good, wholesome food and gets sufficient exercise in the open air. He wants a fine, strong, active child, so he will throw all of his influence on the side of the mother during pregnancy, in an effort to encourage her and help her to keep alive her interest in things outside her own home. He should make her feel that her form is more lovely in his eyes than ever before and try to eliminate from her mind any sense of timidity in appearing public.

The mother naturally looks to the father for protection from material want; he is the provider of the family and stands between it and the outside world. He may have difficulties of his own and many obstacles to surmount and the prospect of increased expenses in the coming of the child may not be particularly encouraging.

Nevertheless, it is the father's business to keep his mind peaceful and harmonious because of his influence on his wife and through her on the child. It is not enough that he looks calm and smiles while in her presence, for if he is worried or uneasy, she will become aware of it.

Strength and courage are attributes which a woman admires in a man and a woman likes to think that her husband exemplifies these virtues. If a man is strong and courageous and stands so in the estimation of his wife, that consciousness within her will be transmitted to her child.

It must be noted, that in modern times women are more independent and follow a career path of their own. Mothers often enough achieve higher duties in their working environment, earn more than the husbands (or partners) and contribute even more in housekeeping and the building of their home.

The contribution of the woman to the family surroundings, her educational standards and the strength of character is very often higher than that of the man. Even so, the support of the husband or partner during the pregnancy and during the rearing of the child cannot be under-estimated.

Of all that has been written, it is obvious that what the mother bears in her inmost character, is passed on to the child and what the mother is, may   be determined to great extend by the father's personality.

The careful nurturing and prolonging of the love between husband and wife, with the cultivation of unselfishness and forbearance on the part of both, will have the effect of giving all the children of their union an equal chance for success and happiness.

Much of the counselling given to the mother, in the preceding chapter, may be followed by the father to the great advantage of the coming child and the comfort of the expectant mother.

THE SNAKE, ON YOUR BIRTHDAY

THE EARTH'S THE SYMBOL
OF CIVILISATION AND REGRESSION
UPON WHICH CERTAIN TIME,
LIVES ARE BORN TO LEARN.

THE SNAKE'S THE SYMBOL
OF THE FIRST FORCE, THE KARMA
THE CAUSE OF LIFE, THE INSTINCT
BORN WITH AND DEVELOP.

WE ARE THE PRODUCT
OF OUR INSTINCT-SNAKE,
OF THE EROS-LIBIDO, STRIVING
PERFECTION TO FIND.

LET THIS AND OTHER,
MANY BIRTHDAYS, FULFILL
YOUR ADORABLE MATERNAL LOVE
AND YOUTH TO SHINE ON YOU.

MANY YEARS AGO

ON THIS DAY YOU WERE BORN

AND SINCE THEN YOU'VE SEEN AND DONE THINGS

OF WHICH I'M PROUD TO THINK.

FOR EVER, UPON THIS DAY,

I PROMISE TO KEEP YOU GAY

AND STARTING FROM THIS DAWN,

LET NEW STRONG LOVE BE REBORN.

VALUABLE GOLDEN SILK

THE SNAKE KEEP

ON YOUR WRIST, TO MAKE YOU FEEL

"A VERY HAPPY BIRTHDAY" - KEEP.

END.

ANDREAS SOFRONIOU'S ANTHOLOGY, ISBN: 0 9527253 0 4.

8. THE DAY OF BIRTH

The mother may have dreamed of a day of opportunity when she will be able to do something worthwhile, placing her name in the annals of those who accomplished great things. The day of the birth of her child is a day of the greatest opportunity that could come to anyone. An inventor may come on a day when he concludes his final experiments and brings into manifestation an invention that will be of untold value to humanity. This sinks into insignificance compared to the opportunity of the mother on the day her child is brought into the world.

She has a chance of giving birth to a perfect baby. Sentiments are expressed in the celebration of Mother's Day, every year, but the mother's greatest day is the natal day. This is the day she brings into life her own child.

Much of the difficulty and suffering which women experience in giving birth to children is caused by wrong habits of thought lodged in the subconscious mind. Through the generations the race thought has connected with childbirth with terrific suffering and many tragic happenings, so the approach to childbirth is filled with fear and read. This fear of pain and trouble causes her body to become tense when it should be completely relaxed and at ease. This tension is one thing that produces pain in the delivery of the child.

We build into our bodies the habitual thoughts which predominate in the subconscious mind. Months of faith and good cheer, preceding the day of birth, the natal day, will lessen the difficulty for the mother, as well as insuring the future of the child. The mother must always remember that any thoughts of fear or dread are destructive and can only harm, while thoughts of confidence and love are constructive and uplifting.

So much has been said about pre-natal influence that the mother may feel that the greater part of her work has been done, so far as her influence over the child concerned. But in reality, her work has just begun. The relationship between mother and child is always a very close one and makes it possible for the mother to guide and train her child as no other person can.

Second only in importance to the nine months of gestation are the first eighteen months of the child's life. In the past it has been thought that the physical needs of the child were the only thing of importance in infancy. Since the child's mental faculties are not developed for several years after birth, it has been thought that the matter of environment and association during this early period has nothing whatever to do with the future of the child.

The mother takes her child as a trust of great responsibility and her work is to guard, guide and direct it so that the end of its existence on earth, it will have given a good account of itself. Regardless of her circumstances, the child during its very early life is her job. The importance of this work is second to nothing. The mother's task carries with it unlimited possibilities.

The inventor who serves mankind through contributing to material comfort and advancement, the explorer who finds new territories for man's home, artists who in pictures, song and story reveal man's deepest thoughts and longings, all of these render a mighty service to the race. But the mother's work is superior to all of these, for within her hands is the shaping of a human life which has within its power the redemption of the world.

A few mothers consecrated to the sacred service of motherhood, willing to give their all during the early years of their child's life, could within one generation, start the world on its climb to greater heights.

The natal day was spoken of as the 'day of pains' through which the mother must pass and sometimes the mother felt gloomy, as to make this experience very perilous. But it can be a day of joy, wondrous possibilities, fulfilment of a mother's dream. With her baby in her arms, the mother experiences peace and fulfilment. An overflowing feeling, that her 'cup of joy' is full.

The baby lying in its cot, prepared for it by loving hands, is feeling upon its small body for the first time the influences from the whole wide universe. It is beginning to use its five senses and in due time, through many trials, will learn with the assistance of the mother to walk upright, speak and use its other functions.

The little body and mind are very sensitive to all influences for as yet all its parts and organs are tender. The mother supplies its first food and causes its first reaction to pleasure. During the first six weeks of the baby's life, its primary urge is hunger. In the natural working out events, this desire is satisfied by drawing the life-giving milk from its mother's breast.

It is thought sometimes that a child must be three or four years of age to be taught anything of importance. Actually, the education of the child begins before birth and every minute after birth there is a struggle in the little body for growth and progress. Instinctively, it tries to use one set of muscles after the other. In reality, it begins to use these muscles even before birth, but after birth these motions begin to have purpose and desire shown.

The baby likes to be undressed and will laugh and coo with pleasure for it feels more freedom and can move its body with greater ease. It is scarcely ever really still, except when sleeping. Soon it learns to hold up its head a little as the back muscles are strengthened.

The baby should be fed at regular intervals and as quickly as possible regular habits established. The child has but one way of expressing its desires and that is by crying. Whether these desires should be complied with or not as part of post-natal training, he should be taught early in life that he cannot have his wants gratified by crying. We do not get what we want by crying for it, but we get what we want because we earned it and have attracted it to us. The sooner the baby learns a part of this important life's lesson, the better it will be for him.

The child should be bathed frequently. Great care should be taken to keep the rectum and sexual parts clean and free from irritation, with just as little handling as possible. Sometimes tight clothing causes the child to touch and handle himself.

At this stage, the mother should eat plenty of good, wholesome food which will bring milk in abundance for the baby. When nursing the child, the mother should be very attentive and communicative with the child. Whilst breast feeding the mother under no circumstances should feed the baby while in an undesirable emotional state, such as anger, hate, fear or worry. It is a well-known fact that if the mother is in a state of anger while nursing her child, her milk is likely to be unfit and the child may become ill.

The mother should keep herself healthy, exercising in the open air and taking some rest everyday. She should not allow herself to become nervous. As she values the well-being of her child, she will maintain her poise and calmness. Try at all times to feel peaceful and contented.

As the child grows in years, he must have outside influences and interests and play with children of his own age. One of the first demands of the child is for food. It is hungry and must be satisfied. There is but one natural food for the infant and that is the mother's milk. Nothing else can take its place, if the mother has properly guarded her own health and physical condition during the pregnancy.

Happy is that mother who is willing and glad to assume the responsibility and nurse her child at her breast. It is difficult to understand how such a natural and eminently satisfactory method of rearing children came into such disrepute and how mothers could be willing to substitute other prepared foods.

There are instances where the mother for one reason or another is unable to supply the quality and quantity of milk which the child requires and so is forced to rely on other means of feeding her baby. Such mothers must do the best they can under the circumstances. The normal mother is well able to take care of her child in the matter of nursing it and nothing should induce her to do otherwise.

It may be that this personal care of the child will interfere in some respects with her social life, or with her business or professional career. If so, she should weigh the child and the considerations in the balance and determine which is of greater importance. No position in society and no standing or achievement in business or professional world is worth, even a small part of the value of one child. Therefore, the mother should gladly relinquish every ambition, for the first part of the child's rearing.

Men may play the financial game and make their millions and it can be wiped out in one stock market crash. They may win fame and fortunes, but other greater achievements soon follow theirs and what they have done is forgotten. They may move the multitude with their oratory, but the world moves on and the idea which they stood may be lost sight of. These are toiling more or less with transient things, necessary, but only steps in the world's advancement.

The mother's task is the directing of a soul, which for a few short years has been given into her care. No human mind is great enough to foretell the possibilities of that soul if the mother does her work well. Can she relinquish one iota of her opportunity to weld all of her influence into a mighty power for the good of her child?

The mother loves her child, with love like no other human emotion. She will deny herself food, clothing and pleasure that the child may not want; she will care for it through weeks of illness without a thought of herself; she will even sacrifice her life to protect her child. In her feelings for her offspring she combines all the elements of the love of an animal for its young, the higher and more understanding love of the human being.

As she holds her child in her arms and feeds it at her breast, she is giving it her love and part of herself. This act binds the mother and child still closer together and may be used as an instrument of great power by the mother.

During the months when the mother is nursing the child, she must realise that what she is in her life and personality, she is imparting to the child. Not only what she is thinking and feeling at the nursing periods, but all of the time during those months.

Psychologists tell us that all mental states are followed by bodily changes and that all consciousness leads to action. This is true of desires, or emotions, of pleasures and pains and even such seemingly non-impulsive states as sensations and ideas. It is true of the entire range of our mental life. The bodily effects in question are of course not limited to the voluntary muscles, but consist in large part of less patent changes in the action of heart, lungs, stomach and other viscera, in the calibre of blood vessels and the secretion of glands.

Since the above is true, the mother should be most careful of her mental health. She should avoid anything which is negative or discordant. Many children come into the world, poor and continue in that direction, because the mother has a sense of lack and poverty. She has permitted her mind to surrender to the appearance of poverty which may surround her at that time. One may have nothing in worldly goods and yet not to be poor; she/he may be happy, contented and fell that he/she has everything.

All negative and destructive thoughts of the mother bear fruit in the life of the child. Later in life, if undesirable conditions manifest, the mother may not realise that she may in part be responsible.

In many respects post-natal culture is as important and necessary as pre-natal culture. If a child has not had the proper pre-natal influence, much can be done for him after birth through ideal post-natal suggestion.

The mind of the child is very impressionable. As the years go by, these impressions become fixations. Therefore, a child may be moulded for good or bad, for greatness or mediocrity. How important then, that parents, teachers, preachers and doctors understand this and register only good impressions. Principles, ideals and other characteristics may or may not become active in the early years, but they are never lost.

THE OLIVE TREE

THE OLIVE TREE GOT JEALOUS

OF MY PRODUCT'S COLOUR

AND THE OLIVES THEMSELVES

MENTIONED THE SWEETNESS

MY OILY OLIVE, NATURALLY GOT.

HOW PROUD IT MAKES ME TO SEE

THAT MY TRAVEL OVER THE SEA

GOT FRUITFUL AND PROSPEROUS,

THANKS TO MY DAYDREAMS

AND DECISIONS FOR EVER

TO KEEP PROSPEROUS, FOR EVER

TO LOVE AND BE LOVED.

WHOM DO I HAVE TO THANK?

BUT MY MOTHERLAND

AND THE MOTHERLAND

I CONJUCTED AND ADORE,

WHOSE DREAMS AND ACTIONS

WERE OF HARD A WORK AS MINE!

I LOOK AT HIM AND SEE

THE BLUE SKY IN HIS EYES

AND THE BLUE RESTLESS SEA

WHICH AMBITIOUSLY, PLAYFULLY

MOVES HIS TINY LEAVES AND PETALS.

EVERY MOVE AND LAUGH

IS A PLEASURE FOR US.

AND EVERY CRY AND SCREAM

IS HIS AND OUR BIG DREAM;

THE FUTURE HEIGHTS AND DEPTHS

CONQUERED, TO SEE

AND HIS SUNFUL, BRIGHT MIND

UPON NATURE, BEAM!

END.

ANDREAS SOFRONIOU'S ANTHOLOGY, ISBN: 0 9527253 0 4

9. CARE FOR THE NEW-BORN CHILD

The arrival of a baby is an event in any home. It is an unusual family that does not become engrossed. This is where a word of warning is necessary. The fact that the baby is sweet and dear makes it difficult for the parents and others to keep from fondling it all the time. Some babies are treated as though they are special playthings for the rest of the family.

The baby should never be on display. It may be seen by callers for a moment in its own room but not disturbed or handled in any way by visitors. During the first weeks and months of the child's life, the greatest need is for peace, quiet and love. These the mother will supply with loving assistance of the father.

The parents should not rush and hurry about the child, shake it or jerk it roughly. There are times of course when the parents will want to kiss and cuddle the child. If for any reason the mother does not nurse the child, she should by all means feed it herself.

This does not mean that the mother must give up outside interests and remain by the child's side all the time. If the child is well trained from birth, it will be content and quiet by itself whether awake or asleep and with someone to watch it occasionally to see that all is well. The mother may get away for short periods and thus keep her contact with the outside world. She should keep any avocation which she enjoys and which enables her to express herself.

It may seem that the almost constant care of the child, which is outlined as her task, would prevent her doing anything but watching it. The mother should never permit the child to demand her presence all the time. She should train her baby to be alone certain parts of the day and by resolutely remaining out of the child's room establish that habit, of the child learning to be alone, occasionally.

This will allow the mother to carry on her own affairs and not to neglect the rest of the family on account of the newcomer.

By now, the mother deals with the child in an intelligent, conscientious manner with certain well-defined objectives in mind. In working out these objectives, the feeding time brings one of the greatest opportunities. In the first place, there must be regular periods for feeding the child for two reasons; First, regular feeding will ad much to the development of proper digestive action and second, the child will begin to form habits of regular, orderly action in life.

Before feeding the child, the mother should sit down quietly and collect her thoughts. The mother must bring herself to this time with nothing else in mind except the welfare of her child.

She will hold the child carefully, gently and lovingly in her arms. The holding and the thought of this embrace will have a tendency to express her love. During this period she will give all her attention to the child. She will talk to the child while it is feeding.

These points may seem strange and unusual to the person who is not a trained child psychologist, but those familiar with psychological principles will know that such loving actions become effective in the life and personality of the child.

If for any reason the mother has to bottle feed the child, the same principle are followed, as those of breast feeding; holding the child in her arm and keeping the bottle next to her chest, with full attention to the baby and together in solitude.

Some may think that keeping the child by itself, or being alone with the mother during feeding, will tend to make it timid and unsociable in its nature later on. No such feeling need be entertained for the reason that this will only be practised in the very early infancy of the child. As the baby gets more accustomed to his/her new environment and adjusts itself to new surroundings, then the child may play with other children and in every respect lead the life of a normal healthy child.

At this stage of world's advancement, it is not possible to train entire families, or communities into the correct methods of caring for the new born child in a way that will ensure its successful progress through life. Any training must seek to develop the individual child.

There are cases where the parents had insufficient knowledge of parenting and the lives of boys and girls were ruined and their chances to become successful destroyed. Parents must learn to let the child go - become free, gradually inter-mix with groups of other children. Nothing is more pitiful than to see a grown man or woman dominated by a mother who refuses to realise that the son or daughter is a child no longer and must assume his or her place of responsibility in the world.

If the mother is to be able to grant the child freedom when he is grown, she must begin at the cradle. It must be realised as early as possible that the infant is not a possession, rather an individual human being with a purpose in life, which may be different and far removed from anything of which the parents can possible conceive. If the child is to be prepared for the next generation, this means that the child needs a more advanced and broader outlook in life.

The love for the child will doubtless be the consuming passion of the parents' life. Their love must be used to bring him/her freedom and open wide the door for him to pass on to great achievement. The child may be destined to deal with affairs and conditions which are beyond the conception of anyone living today.

The world today needs leaders of worth and courage. Imagine what the child may be doing in, say 25 years. It is up to the parents to start thinking of the education and training of the child, now. It is for the parents to prepare for their sons and daughters to be the leaders of the world's progress tomorrow. This is the parents' task. Not a paltry thing of rearing a child to be a citizen of the world, who will bring all of his influence to bear in the cause of right and purpose.

All of the ills and disturbances from which humanity suffers are the result of low ideals and lack of understanding on the part of the individuals, for humanity is made up of individuals.

If the parents of this generation can redeem humanity from the present degraded and materialistic conception of life, then it will be possible to develop mankind to bring about peace on earth and good will among men.

DESTINATION

WANDERING, STRUGGLING, SEARCHING

IN THE ENDLESS DARK SHEET OF THE NIGHT

SCARED OF THE TREE-BRANCHES, BENDING

DRAGONS TO CATCH AND EAT ME.

TO BE BLIND, OR MY EYES FORCED TO CLOSE

PERHAPS BY THE HAND WILL BE TAKEN, LED.

TO STRETCH MY AUDITION, LISTEN

PERHAPS A VOICE WILL CALL, OR A SMELL TO FOLLOW.

EMPTY VESSELS HORRIFYINGLY SOUND.

PEOPLE, SO LOUDLY SPEAK THAT TURNED ME DEAF.

BY LYING DOWN TO REST, PERHAPS TO FORGET,

EXHAUSTED, SUNK IN PERSPIRATION, I SLEEP.

UNBELIEVABLE, MYSTERIOUS, EXOTIC.

MY EYEBALLS ARE TURNING INSIDE, TO SEE

MY SKULL FULL OF SUNSHINE, BRIGHT

LIGHT WITHIN, TO FOLLOW.

END.

ANDREAS SOFRONIOU'S ANTHOLOGY, ISBN: 0 9527253 0 4.

10. THE POWER OF WORDS

The conversation with the child should not be in baby language. The family of the child must remember that the subconscious mind of the child is accepting what is said and the manner in which it is said can be a strong impression on the child.

Parents may think that the kind of language used in the presence of a baby does not matter because the baby does not understand. It is true, its conscious mind does not comprehend, but the feelings which produce any type of conversation are registered on the child's subconscious mind. The parents should know that they must be careful in speech and conduct in the presence of the baby, as they would be in the presence of an older child.

From the time of birth, the parents should choose the very best and purest language. They should speak the words clearly and enlarge the vocabulary. If the parents speak kindly words to the child, they will find a little later, that the child will be addressing his playmates in a similar expression.

The child should be given a dignified name and called by that name. It should not be called 'baby', or nicknamed. It should be called by its proper name. From the very beginning the child should be made to realise that it is an individual human being and of importance in the world; that is it has a name the same as grown ups and is entitled to be treated with consideration and understanding.

It is never too early to train the child. Training, however, should be almost entirely by example and by word. Where children are unruly and bad - disobedient, it is usually because the parents are weak and negative in their inter-relationship with the child.

When the child needs to talk to somebody, he will turn to the parent who has always shown good judgement and steadfastness. He will feel that the parent can help him through his time of difficulty.

For the parents to be firm and positive does not mean that they will use their strength to break the will of the child. There will be little difficulty along these lines, if the parents from the very beginning are constructive in their training, wise in their guidance and good in the use of their vocabulary.

The parents should very clearly make a companion of the child, talking to it intelligently, as though it were an older person and could understand everything they are saying.

They should speak perfect English (as much as they can), speaking slowly, distinctly and clearly.

The parents should talk to the child about many things, including discussions about their own family, other children, places, the world, their professions, the arts, technology, religion and philosophy. Above all the parents must remember to answer the child's questions truthfully.

Parents must agree with each other about subjects of discussion in the presence of the child. The time to settle a difference of opinion is when the parents are absolutely alone. If parents take a given position on some question and later on the child discovers that they have changed their opinion, it may then be told that after further careful consideration, they decided on a change of program due to further factual evidence.

New and different problems in the relationship between the parents and the child will be coming up constantly as the baby days are ended and the child develops and begins to form outside contacts.

There may be fear on the part of the parents that some harm may come to the child from his association with other children, that the ideals they have tried to establish in its mind may be destroyed by children who have not had the advantage of wise training. But if the parents allow themselves to have any anxiety for the child, they are doing it a great harm, for the child will sense such feelings, although he will not understand the cause, but that feeling of fear of the part of parents may weaken and undermine what they have been working to build.

When the parents send the child out to play with other children, or to go to school, they should be doing so with gladness. They should remember that the child must gain a series of experiences in order to develop normally and naturally. It will not be rational to stand between the child and any experiences which come in the natural and normal way and which may result in a warmer character, than otherwise have been developed.

There are good and undesirable experiences and the parents must bear in mind that this is the way life goes on. The parents would, very early teach the child that whatever comes to him in life, must be experienced and learn out of it. The child will soon establish a feeling of responsibility and a sense of security with reference to the experiences in life and whatever learned for his own future. The child will learn by his mistakes and errors and parents can be there to listen and guide.

If the child makes a mistake, serious enough to call for parental punishment, this many times fails to correct the child, because it has no relationship to the error. Any punishment to be effective, must be addressing the cause for the misconduct, made to fit the misdeed. Unjust punishment should be avoided

It is the custom in some families for the mother to make a list of offences committed by the children during the day and when the father comes home from work, to make her report to him and request that he punish the children. This is an injustice to both, the father and the children. Had the mother been firm, it would have not been necessary for her to relegate to someone else the task of inducing obedience. Every act of disobedience and tendency to violate, should be dealt with as it comes and settled at that time.

MY TREE

IN THESE DEEP DARK SURROUNDINGS,

I AM FOUND AMONG LOFTY WALLS

TRYING TO REACH THE TOP, TO SEE THE GLITTER

THE SUN'S, COMING AND PASSING OVER.

WITH TIREDNESS, MY UNWANTED FRIEND,

ACCOMPANING MY MELANCHOLIC DROWSY EXPRESSION

I STARE AT THE.., WHAT I REALISE NOW,

CURIOSITY OF WHAT IS ON

OTHER PLACES I HAVE NEVER SEEN BEFORE.

IF I COULD ONLY GROW IT HIGH

ENOUGH TO CLIMB ON IT, COULD I BUT LEARN.

PUT ITS HEAD INTERCOURSING THE SOIL DEEPLY

AND IN TIME, ON ITS HANDS AND LEGS SIT..,

TREE, I SHALL MAKE YOU BIG AND STRONG.

THIS SENSATIONAL DARKNESS DROVE ME MUTE

I PROJECT WITHOUT AN EXCUSE

WITHOUT TO BELOOK, BETHINK... SO LETHARGIC.

IDLENESS... GO AWAY, I'LL STOP LISTENING -

TO YOU RETROSPECTION AND INTROSPECTION I COME
AGAIN

TO BE SHOWN WHAT I HAVE ALREADY GOT.

I HAVE IT, IT'S TRUE, THE TREE IS THERE

USED TO ALL UPS, DOWNS AND THUNDERSTORMS,

READY TO FACE EXPERIENCES OF ANY SORT.

SO LONG IT TOOK ME TO CLIMB YOU,

BUT HOW EASILY AND CONFIDENTLY.

AS YOU ARE MY OWN TREE, I SIT

ON YOUR TOP AND ADMIRE THE BEAUTY,

GODDESS OF THE RESOLUTIONISTS, DETRIMENTED,

THOSE WHO ATE THE SWEET POISON

LIFE, SPOON BY SPOON TO THE BOTTOM.

GODS AND GODDESSES OF ALL THE OBJECTS,

SUBJECTS TO THE APOLLONIAN DESIRES,

GIVE ME THE CHANCE TO ANGELISE OTHER TREES

STRONG, TO DEVELOP THE NEW GENERATION

TO PUSH UP AND BREAK NEWLY-FOUND SHELLS.

END.

ANDREAS SOFRONIOU'S ANTHOLOGY, ISBN: 0 9527253 0 4.

11. ANSWERING QUESTIONS

Very early, the child begins to ask questions and naturally he runs to the mother for an explanation. The mother many times has the answers ready, perhaps before the questions are even asked. Even so, many times she will be surprised by some unexpected queries. The main thing is to answer intelligently and honestly.

One of the very important the child will ask is the subject of sex. Some children are interested in sex much sooner than others. It is common for children as early as in the fourth year to ask such questions and sometimes even younger. The parents may be speaking of some place where they were before the baby came and the child may want to know if he were there as well. When told that he was not there, he immediately wants to know where he was.

A new baby may come into the family and the very sensible question is asked by the four-year-old as to where the baby came from. The parents have such a wonderful opportunity to impact one of the truths of life in a way that would satisfy the inquiring mind and protect him of wrong information from boys and girls in a vulgar way.

The parents must explain that every living thing has a father and a mother. This knowledge can be unfolded in a simple language, the wonders of the animal world and how the birds mate, build their nest and hatch their young. The truth will charm the child more than a fairy tale.

Many times it is more expedient to tell the child that a baby is coming and give him an opportunity to show interest in this great family event. Give him some information on the coming of the baby. There will be no awkward questions after the baby arrives and he will be on the solid footing of confidence and understanding with his parents.

When the child approaches a parent with questions about life, the parents can explain in simple language which he can understand, that every life, whether plant or animal, has the power of reproduction. The child will be impressed with the sacredness of life, the purity of sex and the wonder of parenthood. The parents can explain that the loving union between the mother and father started his life as a tiny seed in her abdomen and that the mother tenderly nurtured him/her there for many months with her own blood, until he was strong enough to live outside her tummy. When a child has accurate information, he will not be interested in the unclean explanation of his companions.

Sometimes, it is asked at what age sex information should be given to children. This depends entirely on the child. It should always begin when the child asks questions regarding sex. Each question asked should be answered truthfully and when the child is satisfied with the answer, let it go at that. If he finds that he gets a truthful answer and he is not told to run out and play, or told not to ask stupid questions, the next time he is puzzled about something, he will run to the parents, who he thinks they know a lot and always tell the truth.

Naturally, not a lot of detail will be given at one time, for the child takes but one step at a time and as he requires further information, he will ask for it, if there is a bond and confidence between him and his parents.

The necessity of beginning early with the child is emphasised, just as soon as there are questions in his mind about sex. If the parents begin then, they will find only innocence and trust in the child. Frankly, the information can be given step by step, as and when required.

The difficulty in the past has been to defer the teaching of sex education until puberty or early maturity. When it is postponed until that late date, its benefit is negligible, if nor entirely lost. Lectures are delivered in Colleges on the dangers of venereal diseases and Aids, the terrible catastrophes which follow and the possible violation of the law of sexual purity. But while these may instil fear into the heart of the young people, they do not prevent wrong doing.

The purpose of sex education is essentially protective. To secure the maximum comprehension of the subject, the foundation teaching ought to start before the young person seeks expression of his/her sex characteristics.

Children are carefully instructed in every other subject and phase of life, except sex. Sex is the most important of all questions to be asked by the child, for its influence in our society affects every other phase, with the power to make a success or a shipwreck of life.

Sometimes it is suggested that the school is the place for sex education. But, a little careful thought will convince any one that the school can support what the family have already explained. In many cases the school cannot handle this subject. In the first place, many children already know much about sex and many have the wrong ideas before entering school. Others may be entirely innocent of any knowledge along this line.

This would preclude giving class or group information and to attempt to give private instructions in the matter of sex to a large number of children, would make it a tremendous task for the teacher. The time a teacher can devote to each child would be negligible, as to accomplish little. Again, it would be a difficult matter for a teacher to approach the matter in a frank way with the children coming from homes of such diversified concepts of the whole problem.

Additionally, the fact that many teachers may have had no adequate education on the subject of sex, it means that they are not qualified to deal with the child on this most important matter.

The medical General Practitioner is, also, handicapped in any instructions he might wish to give to the children of his community. He would have the same obstacles to overcome as the teacher and would have to meet the prejudices and false standards of modesty on the part of many parents.

The school is for the purpose of carrying on the education of the child along lines not easily as satisfactorily handled in the home. But in the matter of sex education, the parents are the natural teachers of the child and this work for them is much easier than for others. It is, therefore, the responsibility of the parents to lay the foundation of moral integrity in the life of their child.

One thing the school can do and which is already doing to some extend, is to give instructions in general hygiene and teach the child the fundamental truths of biology which will enable it to understand easily, the origin of human life.

There will always be many children in school whose parents have given no instruction whatever leading up to an understanding of sex life as manifested in plants and animals. Therefore, such instructions in the school are most essential. By studying plant life, the child may be lead in thought and reasoning up to the point where he will understand something of the law which brings human life into expression.

Another thing schools do to great advantage, is the setting aside of time when parents and/or children can come and listen to discussions on the subject of sex instruction for children from some one who really knows how it should be taught to little children. This person may be a teacher who specialises, or it might be a family doctor who is well-qualified to explain these facts in a simple way which all parents can understand. In fact, the speaker might be a mother who is particularly well-informed and in a position to help other mothers.

It is usually the function of the school to supply the resources. The parent-teacher association might take up some work along this line.

Parents must realise that much of the disease and unhappiness of their children after reaching maturity, results from a lack of instructions in the early years, including the answering of questions on sex.

When a child is in his teens, it is too late to give him the most effective advice regarding the evils which assail him on every side. The time to start is as a baby in his mother's arms, when every breath she may enfold him in an understanding appreciation of the beauty of the creative impulse.

The question may arise with parents, as to how they can train the child in the right way in the face of so many opposing influences, found in the close environment and while away from home. If the parents do their work well in the first five years of the child's life, they will have established themselves firmly in his confidence and be in a position to protect him from most of the harm to be met in his school and social environment.

The child cannot be protected by keeping him at home, away from other children, but there can be such a positive, strong, healthy influence at home, coupled with complete comradeship and confidence between him and his parents, as to practically ensure his safe passage over the perilous journey from infancy to maturity.

Naturally, not all questions by a child can be considered in the space of a book. The most important one of all, the question on sex has been given consideration. Another question which troubles children more than the parents sometimes is the subject God, religion, heaven, etc.

The sensitive mind of a little child is easily impressed with thoughts of fear, doubt and misgivings. In this case, great care should be taken, so that the child's conception of God will be one of fatherhood, motherhood and love. Time should be taken to explain any questions regarding religious things.

Death is another subject which puzzles the child and it is important that very early he has a constructive understanding of it. Otherwise, he may carry throughout life a phobia of death. The answer on death a parent gives depends largely on the individual philosophy, whether a person goes to paradise, or life as we know it ends here on earth, relies totally on the belief of the parent.

If bereavement occurs in the family, of someone very close to the child, it may be possible to handle such a trauma with the help of a counsellor specialising in bereavement cases. It is always wise to seek the help of an individual who is professionally accredited and where empathy can be shown.

In answering the questions of the child, the parents should be patient, sympathetic, and understanding. Like any other question, this question is of very great importance to the child who is trying in his childish way to think through some subject which has come to his attention.

The parent who has time to stop anything she/he may be doing and lead the little mind on in its search for information and knowledge, will later on, find her child when grown to youth or maturity, still coming to her/him with his difficulties and problems. It is the parent's interest through the early years which builds the bridge that carries them together into the future of confidence and companionship.

PARADISE

PARADISE IS,

WHERE THE SUN SETS

AND THE HORIZON ENDS.

WHERE THE BLUE SEA WAIVES BEAT

THE SHORE,

UPON WHICH THE BRIGHT SUN RAYS BEAM.

PARADISE FORMED,

WHERE THE BLUE SKY BOWS OVER,

GENTLY BENDING TO JOIN

THE GREEN GRASS ON THE LAND,

WHERE THE TREES GROW TALL

INTO GIANT-LIKE GENTLE CREATURES

REACHING HIGH TO PROTECT

ALL THINGS, WITHIN THEIR OWN REACH.

PARADISE EXISTS,

FROM THE DEPTH OF THE FOREST

AND FAR INTO THE HILLS,

WHERE LAUGHTER IS CONTAGIOUS

AND LOUD HAPPY VOICES ARE HEARD.

EVERY CHILD PLAYS INNOCENTLY

AND ALL PEOPLE COLLECT THE FRUIT.

THERE, IN THE OVERGROWN GARDENS,

REJOICING PEOPLE, OLD AND YOUNG,

GATHER STORIES TO TELL,

WITH NO CARE IN THE WORLD

LOVING THOUGHTS THEY EXPRESS.

PARADISE ENDS,

WHERE THE ISLAND CURVES, GENTLY

INTO THE CALM, BLUE SEA.

FOR THE HAPPY PEOPLE

KNOW NOTHING ELSE,

BUT THEIR WORLD AS IT IS.

END.

ANDREAS SOFRONIOU'S ANTHOLOGY, ISBN: 0 9527253 0 4.

12. OVERCOMING DIFFICULTIES

Children sometimes are difficult to control and they do not wish to obey their parents. In an effort to protect the child from certain dangers, parents resort to an appeal to fear, to secure compliance with their directions.

Many children are afraid to go around corners, for the bogey man is there, he is told. He must not go off the porch or a policeman will get him; in the dark cellar or closet lurks the monster and so on until the child is beset with a series of fears of every conceivable nature and to these fears can be traced many serious difficulties and failures of later life.

It is quite customary for the parents to read fairy stories to the children. Great caution must be exercised here for, while there are a few fairy stories that may be read with safety to a child, many of them are not desirable. In a fairy story very often, while there is a beautiful character, doing splendid deeds of courage and honour, coming to the rescue of someone in distress, there are other characters portrayed whose actions are evil. In the stories, these evil ones some times carry away little children and hold them captive amidst the most alarming and frightful surroundings. Such a story builds up phobias in the child.

While the child listens to these tales, fascinated, he is filled with fear and terror. When the child goes out in the dark on his own fear and panic fills his heart. Venturing into the garden, he expects to see a hideous monster approaching him, resembling the one he has been listening to.

This fear and horror of the dark, of ghosts and haunted houses become so developed in the child, as to remain even after he grows into manhood.

The child must be taught from infancy that there is nothing in the universe to fear, nothing in the whole wide world which one should blindly be afraid of. By reading bad stories, the parents instil senseless fear into the children. Unwillingly, the parents are transferring the mental life of their children from almost the 21st century, back into the dark ages, where superstition and stupid phobias reigned supreme.

If a child seems to be afraid of certain things, as quite frequently is the case, steps should be taken at once to remove that fear. This cannot be done at once, as it requires time. The child should never be laughed at, nor ridiculed, nor should his fear be made of any great importance. If the child is afraid of the dark room, the parent may accompany the child in the room. On opening the door of the dark room, the parent may explain that the room is dark because the light is not switched on. Turn on the light and the darkness will disappear.

In most cases the face of the child will light up and smile. As and when the opportunity arises discuss the subject of fear with the child and explain in logical sentences, in a matter-of-fact manner and in a calm voice, how natural everything is and how everything can be seen rationally.

A child should never be taught to fear a policeman. He should be taught that policemen are there for the protection of children and if in difficulty at any time, he should go at once to a policeman. Officers are very kind to children; they are parents themselves. If the child is still afraid of policemen, whether this is due to the uniform, or because they heard unsuitable stories, the parent/s should take the child for a walk and visits, which would include a visit to a police station or to meet a policeman on the street.

Familiarising the child with the thing he fears will often banish that fear. Often the fears of the child are but the reflections of the fears of the parent/s. In their feelings the may fear many things and the child who can be very sensitive will express fear of a most unreasonable nature.

Love, confidence, faith and hope are splendid antidotes for fear, doubts and misgivings, If, the parents build their own character with such attributes, the child will automatically identify himself with the parents and become peaceful and poised, with courage and strength.

EROS

SEX, DIRTY AND FILTHY WORD

SWINE'S GREAT FRIEND

WHICH TURNED TO BE

MIXED UP WITH MUD AND PIGS.

EROS, GREEKS' BEST PRAISED FRIEND,

CUPID, ROMANS' IMMORTAL SYMBOL,

LIBIDO, MY OWN ENERGY

MY ID, THE ADORABLE DRIVE.

RELIGION, UNIVERSAL NEUROSIS,

THEOLOGISTS, PERPLEXING AND OBSESSING,

PARENTS AND EDUCATIONISTS RESTRICTING;

THE BEAUTY OF EROS - EVERLASTING.

END.

ANDREAS SOFRONIOU'S ANTHOLOGY, ISBN: 0 9527253 0 4.

13. THE HEALTH OF THE CHILD

The enlightened parent knows that practically all the aches. Pains and illnesses of children may be prevented by wise care and supervision during infancy and that the parents' own attitude will tend to bring about splendid physical health in the child. A strong, healthy child with great vitality will have the power to resist practically every disease which is common to children.

The Nation Health Service still has some of the best services in the world. The parent, by listening to the Health Visitor and by attending the various clinics will find most of the answers and queries they may have. The nurses and the family doctors are only too pleased help with any uncertainties on vaccinations and preventative medicine.

A large percentage of sick children can be saved if parents understand health and refuse to accept appearances as the real facts of life. If in doubt and if further assistance is needed, ask the clinic personnel. They have been trained and most of them have been practising for years. They know how to attend to individual problems, such as the parents may be facing.

On the other hand, it must be remember that a happy environment free from fear, anger, discord and expressing love, gentleness, courage and faith, will act like a tonic on the child and tend to restore to health the one who is ill.

It is not at all necessary that a child be sick and parents should not for one moment believe in sickness. Psychologically, they should only believe in health, think health, live health, talk health and expect health in themselves and in the children. Health can be one of the strongest virtues of their family.

Parents should not show too much concern during a child's illness, nor should the child be unduly petted and pampered and permitted to rule the household just because he is ill. While doing everything possible for him, the matter of his illness should be treated with apparent unconcern and in a matter of fact way so that it may not make too much of an impression on his mind, which might lead to a continuance or recurrence of the experience.

The habit some parents have of producing a thermometer and taking a child's temperature every time his face is flushed, has a very bad effect on the child, for it is a suggestion to him that he is probably ill and that his parents are uneasy concerning him. Nothing could be worse for a child than such attitude.

If parents could avoid all talk concerning sickness and discuss other matters of interest, the children will think little of illnesses. If it is mentioned that there is a so-called epidemic in the town or community, or some child in the neighbourhood is ill, then it is the duty of the parents to take such attitude as will be constructive and helpful to their children. They should never permit the negative side to be emphasised, but state that the child in question has a strong body and will soon recover. They may state that everything will work out for its good. Such conversations in the family circle will have a strong influence on the lives of the children and if the parents talk freely along these lines, their children will come to accept these statements as the truth.

Instead of giving the child a tablet and letting it go at that, the importance of observing a healthy diet, exercise, rest and mental calmness must be observed by the parents and explained to the child. All fears and worries about diseases should be banished. The child will sense this confidence on the part of his parents and respond to it in a wonderful way.

UPBRINGING

THE BRUISED BOTTOMS OF ADOLESCENCE

COULD NOT BEAR THE TOUCH OF THE CHAIR.

BLEEDING NOSE FROM PUNCHES

A PUDDLE OF BLOOD ON THE FLOOR.

TO SPARE THE ROD WOULD SPOIL THE CHILD,

PARENTS INFLICTING PAIN AND WOUNDS

FOR THE CHILD'S OWN GOOD, THROW HIM OUT

STARVE HIM TO SHOW HIM WHAT LIFE IS ALL ABOUT.

FOR THE CHILD'S OWN GOOD, WHO GROWS UP

WITH HATE IN HIS HEART, NO LOVE, NO CARE.

TOUGHENS UP WITH STREETFIGHTING, A GANG LEADER

WITH OWN NORMS, RULES AND LAWS, DIFFERENT VALUES.

END.

ANDREAS SOFRONIOU'S ANTHOLOGY, ISBN: 0 9527253 0 4.

14. Habit Formation and Punishment

Regarding the subject of punishing a child, there has been a great change for the better during the last few decades. Fifty years ago, the average parent regarded the period of six to twelve years as the time to break the child's will - whatever that meant. The cruelty inflicted in the process, even hinted of the tyrants of the Inquisition. During those years, the child does not always choose to obey commands blindly and some times in rebellion, declares that he is not going to do a thing. The old fashioned father in response had his whole being surcharged with the responsibility of fatherhood. He felt the eyes of everybody around him, watching to see what he was going to do when his authority was so challenged.

Imagine the lad, seized and hustled to the shed. The closing of the door spares us the cruel sight that ensued cruelties, which were prolonged until the boy had promised to obey his parents in all matters whatsoever, whenever and wherever. These promises were sweet music to the father's ears as he came out of the shed with all the fatherly pride and self-commendation reserved for heroes on.

He was too blinded with self-congratulation to see that he had done his child a gross injustice and that far down in the heart of the child, he had engendered a hatred that would grow until it would have the power to change and even destroy some of the most beautiful things in life.

When the modern father faces a similar situation, he realises that his child instead of requiring to break its will and rush to enact the shed scene, which will be ashamed of all his life, he tasks his ingenuity for a constructive solution to the problem, which will help the child.

Any talk to the boy, or girl, must be tactfully handled. If it is given in a form of lecture, it will do more harm than good. Somehow, the child must be drawn into the conversation, making interesting enough to lead to questions or an expression of the child's point of view.

Socrates taught his students by asking and answering questions, which is the ideal way of imparting knowledge. If one has sufficient interest in a subject to ask questions, his attention is assured while the answer is being given. Children from eight or ten years on, think much more deeply about life than other people realise. Information and constructive suggestions from the parents, relative to the important facts of life, will take root and bring forth fruit.

The parents should make every effort in the early years of the child's life, to establish a strong friendship between themselves and the child. This is not difficult, if the parents are sincere in their desire to meet the child on ground which is familiar to children. The difficulty usually is that the parents expect the child to look at matters from the adult point of view. This is impossible and may build a barrier between child and parents, which cannot be surmounted unless the parents are willing to concede their viewpoint in favour of the child.

Also, the parents must be willing to give some time to the child aside from providing food, lodging, clothing and schooling for him. The basis of any real help is companionship in which there is respect and confidence in both sides.

Whatever the character of the individual on whom the child depends, that person is the measure of his influence upon the child, whether or not, the parent, guardian or teacher realises this fact. It is highly desirable to direct the child in a certain way and eliminate traumatic habits.

Much of our education is benumbing and instead of developing the thinking abilities of the child, it dulls them. There is no power in the universe like the power of thought.

Its skilful use has brought humanity many social developments, be it political systems, religious beliefs, scientific methodologies and artistic expressions. Parents and teachers should teach children how to think.

By nature, the child is a thinker. This is indicated in many questions it asks. Many of these questions imply that his thinking is based on standards and the truth in all situations. In many cases, teachers and parents dictate silence to the child, especially when they cannot answer the questions raised by the child. Thus, he is denied the right to think and express his thoughts.

What a different world this would be today if, instead of stultifying the child's desire to think for himself in the past centuries, he had been encouraged to think and search for the correct answers to his questions. In that case, the percentage of persons capable of individual, independent thought would probably be about double of what we have now. Definitely more Doctors of Philosophy and numerous research scientists.

Our system of education has produced a society of standardised human beings. This can be seen in the tendency of children and youth to do all things alike, to dress the same way, to talk using same expressions. If these young people had been taught how to think for themselves, they would have individual freedom to think and act - express their individuality. As it is they are following blindly some fragment of a truth, which they have not the ability to properly analyse and apply.

The child can be taught to think for himself, if the parents will take the time and have the patience to talk with him as though he were an intelligent person. This does not mean that the parent is to permit the child to argue about every request made of him, but it is rather the attitude of the parent and teacher toward the child, substituting respectful consideration of the child's conversation, for the usual condescending assumption of the indisputable wisdom of the grownup.

As soon as the child is old enough to express understanding and has reasonable degree of intelligence functioning, it would be helpful to teach him to observe the world around him,. The sense of beauty seen around him will be a never-ending source of pleasure in later life.

The parents must praise the child for every desirable act and deed the child performs. Instead of finding fault with the child, or calling it bad names if it makes a mistake, it should be lovingly shown a better way.

In order to overcome the possibility of developing conceit or undue self-importance in the child, it should be taught that all children are potentially great and that they have within themselves the same possibilities of being fulfilled as their own child has.

Any natural talent which the child seems to have may be emphasised and its development encouraged. Always be sure that the suggestions are suitable to the natural inclination of the child, his aptitude, and interests. Needless to say, the parents may not influence the child with a preference of their own for a particular career.

DILEMMA

SHE PASSED ME BY, HER SMILE TEMPTING

WITH HER BREASTS PRESSED HIGH,

HER LEFT BUTTOCK TREMBLING UP AND THE RIGHT GOING DOWN,

HER WAIST, A RING WITH A TIGHT WIDE BELT

AND HER LONG LEGS STANDING IN HIGH HEELS,

THE GRECIAN COLUMNS OF THE GOLDEN AGE.

HER FACE TURNS BACK TEASINGLY, SHOWING

HER SHOULDER-LONG BLACK HAIR, WITH THE BREEZE

HER DARK EYES PENETRATING SEARCHING TO SEE

IF I AM TEASED ENOUGH, FROZEN OR PROMPTED.

PROVOKED, I FOLLOW

THE LONG LEGS,

THE PROTRUDING CHEST,

THE WELL-FORMED BACKSIDE,

HER TRAINED SMILE

AND HER WAVY SILKY HAIR.

I WONDER, DO I

TOUCH THE BEAUTY, FEEL FROM AFAR, OR TURN BACK?

END.

ANDREAS SOFRONIOU'S ANTHOLOGY, ISBN: 0 9527253 0 4.

15. THE PERIOD OF TRANSITION

The adolescent period is that part of life that is known as the stage when one is growing into maturity, the 'teens'. During this period of life the adolescent is not a child and yet he is not grown up. The seven years of adolescence play a very important part in life. For during these years, the personality, habits, mannerisms, likes and dislikes mature and they are usually carried on through life.

Those responsible for the training of the youth would do well to guard this period as though it were a rare and precious jewel. Parents and teachers should exert every effort to make certain that the adolescent is properly developed during the momentous era of life. The importance of these seven years is often taken too lightly by the ones who are directly responsible for the future welfare of the child.

 parent who reads this book and has a teenage child, may feel that not having given the child the care and attention, as mentioned in the earlier chapters of this book, his opportunities for being of great help to his child is past. This not true, for there is still much that can be done. In fact, it is never too late to do one's best. The parent should begin from the present, knowing that his child has imperative need of him through the period of adolescence.

This is a period of great importance from the physical standpoint, for radical bodily changes are taking place. Particular physical care and direction are required in order that the youth may pass this critical stage with his body unimpaired. Both, the boy and the girl are subject to these changes and apart from nourishing food, study and companionship, they also must have information regarding their physical make-up, the reason for the change through which they are under-going and the effect of this change on their future.

If the parents have answered candidly and honestly the questions asked earlier in the child's development (early childhood), the child will already know most of the facts necessary for the early adolescent stage. If these facts are not already known to the teenager through conversation with the parents, they should be explained at the first chance given. Otherwise the young person may be unduly alarmed.

The habit of masturbation is sometimes formed at this time. Parents should not take this seriously, for according to some authorities on the subject, it is a habit quite common among boys and girls. It is nor harmful as was formerly thought.

No matter how timid one may be about sex, parents must not shirk their duty to the child. It is a most important subject and the incorrect understanding may spoil the child's future. There are cases where the parent feels that he is not sufficiently informed or does not know how to impact the details, he should find out about the subject as much as possible and then skilfully pass on the information required.

Sex is usually thought of with reference to the reproductive system of the species only, but that is only one of the phases. It is the creative force in the body and its conservation and right use will enable the man or woman to create in the fields of art, literature, invention, or in whatever direction his/her talents may lead. The most gifted people are the most strongly sexed. The human race certainly finds sex pleasurable and any normal couple will find a lot of pleasure in making love.

At this point in life, in adolescence, the mind must be developed and reason must be exercised. The adolescent demands larger fields of conquest in knowledge, because it is natural to be endowed with larger capacity for reflection. New desires enter his life; he becomes emotional, imagines things and lives a life of dreams and illusions.

With the arrival of puberty and the adolescent period comes a certain degree of haste in the development of the mind. The mind now develops much more rapidly. It demands new and broader fields of knowledge. The child is now leaving behind it the weakness of childhood and is gradually preparing to assume the mental strength of maturity.

The teenager is likely to develop a liking for light fiction, such as detective plots and love stories. As the education becomes more demanding the youth graduate to more serious studies and very often they only become interested in text books. Some topics of ephemeral interest may be useful. Parents can make it a point of discussion at the dinner table. A variety of literary interests can be talked about.

From fourteen to eighteen years of age, is a very difficult time for the parents to handle the young person diplomatically and constructively. The parent must be very careful in his dealing with the child. Always discuss the whole situation freely and remember that the child does not exclusively belong to the parents.

The parents should try to feel that even when a son or daughter leaves home for a far away country or even close by, this is part of the young person's learning. Be around when they call on you, listen to their progress or regression and be prepared to share the experiences, whatever they may be. Remember, as parents, you should trust those you brought up, the people you trained and cared for, even before they were born, from the moment of conception.

The enlightened modern parents, when the offspring has reached this stage of development, must have already talked earnestly with him. The parents have noted the gradual lessening of holding onto the child, as the child grows in years. In their discussions with the child they incorporated the importance of education, of growing up strong and independent and of things to be encountered.

Now with confidence in the young man or woman, the parents allow for the leaving of home. Perhaps embarking onto a to a university study, another place for work, or travelling until he or she is ready to take on the life of an adult. It is not easy for parents. They will feel the empty corner of the home. It will for sure be, at times, a lonely life for the young ones, a struggle until they experience whatever life has to offer.

Neither the parents, nor the young people must forget to keep in touch with each other, to continue their discussions and sharing of their love for each other.

These emotions never go away from the individual. Exchange information with each other, as often as needed. A life was shared for many years, this cannot be forgotten. On the contrary, with all the love and care given and taken, now is the time to look at the beauty and confidently admire the maturity of the next generation. With such a joyful parenting the young people who left home can but contribute to the betterment of the next generation - the improvement of society at large.

ROUTINE

THE STREETS ENDLESSLY WIND
INBETWEEN HOUSES, HIGH AND LOW.
BUILDINGS REACHING THE SKY
AND HOMES CLOSE TO THE GROUND.

VOICES ESCAPING THROUGH THE WINDOWS,
CHILDREN RUNNING FROM ROOM TO ROOM,
MOTHERS CALLING FROM AFAR;
THE GAMES DROWNING THE SOUND.

THE COMPUTERS HUMMING IN THE OPEN SPACE
OF THE OFFICES, FROM FLOOR TO FLOOR.
FATHERS ABSORBED IN THEIR THOUGHTS,
FACING DOWN AT THEIR DESKS.

FATHERS LONGING FOR THEIR WIVES,
MOTHERS PREPARING FOR THEIR HUSBANDS,
CHILDREN SECURE AT THEIR SCHOOLS,
TEACHERS LOOKING FROM AFAR.

COME WEEKEND, IN THE FRONT ROOM
WATCHING THE SET AND DAYDREAMING,
FAMILIES SITTING CLOSE, TOGETHER.
EMPTY OFFICES, UNTIL THE DUST SETTLES.

END.

FROM ANDREAS SOFRONIOU'S ANTHOLOGY, ISBN: 0 9527253 0 4.

16. THE RESPONSIBILITIES REVIEWED

In the contents of this book one may have noticed the gradual lessening of the directing hand of the parent as the child increased in years. Before birth the mother's control an influence are supreme and during the first eighteen months the mother shaped the character of her child. Up to five years age her power is great. As the child gains in knowledge and understanding, gradually the parental guidance recedes and the child increasingly relies on his/her own resources.

So, all the way along in the life of the child, the parents encouraged the child to think for himself and to make his own decisions as far as compatible with the good of all concerned. Little by little the parents receded from a position of great prominence in the direction of the child.

As much as possible, all through teenage-hood and thereafter maturity, the control of the child was based on friendship and companionship.

It is a hopeful sign that the parent of today takes an altogether different attitude toward his grown children than prevailed in some families a generation or so ago. The parent today gets his greatest happiness in having the children go into the world and be successful. The greatest relationship is the one between the parents and the children. It has the possibility of lasting longer and the chance of attaining greater beauty.

It may be observed that the authority and responsibility of the parents have been gradually diminishing throughout the adolescent period and when maturity reached the child, the responsibility and authority of the father and mother ceases absolutely. Just as the child has been finding it difficult to adjust itself to the gradual increase of responsibility, the parents will find difficulty in gradually withdrawing responsibility, but it must be done. For when children reach maturity, the parents' responsibilities cease completely.

In reaching maturity, the child becomes strong in mind and body, truthful to his fellow men, starts a career and a new relationship. Parents, naturally feel just a twinge of jealousy. Simply, the parents will miss the little things they used to do for the child. But in time, gradually and with wisdom they let go, for the child is now a mature person, with freedom to choose his/her life and inter-relationships. They marry, they share things with partners and with lots of new experiences, they develop even more.

As parents, you have now performed your duty in establishing the proper relationship between you and your children. The loving hand of parenthood has been ever ready to lead and direct in the right way. Your children have arrived at maturity, and they have gone out on the pathway of life for themselves. You have done your duty and you believe that everything will work out all right and that your children will be happy and successful.

THE BETTER HALF

IMAGINE,
TWO HEADS, FOUR HANDS, FOUR LEGS
ON A HUGE BIG BODY
ALL EXTRIMITIES FOR EVER, TOGETHER AS ONE.

INTELLIGENCE,
DOUBLE THE ABILITIES AND STRENGTH,
FASTER THE TASKS COMPLETED,
QUICKER THE DISTANCES COVERED.

JEALOUSY,
THE GODS FELT
THE CREATURES COMPETING WITH THEM,
FRICTION AMONG THE HEAVENLY AND EARTH BEINGS.

IN A HURRY,
THE GODS SEPARATE THE ANIMALS, CUT THEM IN HALVES.
SCATTERED THE HALVES, AWAY FROM EACH OTHER,
IN PUNISHMENT.

FOR EVER,
THE HALF-CREATURE
ON TWO LEGS, WITH TWO HANDS AND ONE HEAD
SEARCHES FOR THE OTHER - BETTER HALF.

END.

ANDREAS SOFRONIOU'S ANTHOLOGY, ISBN: 0 9527253 0 4.

Books publications by Andreas Sofroniou include: *Information Technology, Management, Psychology, Poetry, Philosophy, Epistemology, Fiction, and Medical Sciences.*

Andreas Sofroniou DSc PhD ExecMBA FIoD is a retired Consultant Psychologist, an Information Technology Executive with international organisations, a Principal Adviser to Government Departments, and an affiliated Professor & Research Fellow of various Institutions & Universities in the USA and the UK.

Synopsis

The book concentrates on the upbringing of children and offers guidance in establishing the right relationship between the child and the parents. It deals with the pre-natal and post-natal influences and expands into the realms of continuous development of the human personality. Remembering that human personality with all its complex characteristics never stops developing; from the foetus stage, to birth, growing up and to dying in old age.

.